SpringerBriefs in Complexity

SpringerBriefs in Complexity are a series of slim high-quality publications encompassing the entire spectrum of complex systems science and technology. Featuring compact volumes of 50 to 125 pages (approximately 20,000–45,000), Briefs are shorter than a conventional book but longer than a journal article. Thus Briefs serve as timely, concise tools for students, researchers, and professionals.

Typical texts for publication might include:

- A snapshot review of the current state of a hot or emerging field
- A concise introduction to core concepts that students must understand in order to make independent contributions
- An extended research report giving more details and discussion than is possible in a conventional journal article,
- A manual describing underlying principles and best practices for an experimental or computational technique
- An essay exploring new ideas broader topics such as science and society

Briefs allow authors to present their ideas and readers to absorb them with minimal time investment. Briefs are published as part of Springer's eBook collection, with millions of users worldwide. In addition, Briefs are available, just like books, for individual print and electronic purchase. Briefs are characterized by fast, global electronic dissemination, straightforward publishing agreements, easy-to-use manuscript preparation and formatting guidelines, and expedited production schedules. We aim for publication 8–12 weeks after acceptance.

SpringerBriefs in Complexity are an integral part of the Springer Complexity publishing program. Proposals should be sent to the responsible Springer editors or to a member of the Springer Complexity editorial and program advisory board (springer.com/complexity).

Domenico Lepore · Angela Montgomery ·
Giovanni Siepe · Francesco Siepe

From Silos to Network: A New Kind of Science for Management

A New Kind of Science for Managing Complexity

Domenico Lepore
Intelligent Management
Ottawa, ON, Canada

Angela Montgomery
Intelligent Management
Ottawa, ON, Canada

Giovanni Siepe
Intelligent Management
Toronto, ON, Canada

Francesco Siepe
Intelligent Management
Baronissi, Italy

ISSN 2191-5326 ISSN 2191-5334 (electronic)
SpringerBriefs in Complexity
ISBN 978-3-031-40227-2 ISBN 978-3-031-40228-9 (eBook)
https://doi.org/10.1007/978-3-031-40228-9

This Springer imprint is published by the registered company Springer Nature Switzerland AG
The registered company address is: Gewerbestrasse 11, 6330 Cham, Switzerland

We dedicate this book to our parents.

To Annamaria and Giulia, your unconditional love and support makes us better, stronger and more capable of bringing our ideas forward.

To Agnes, Harry, Francesco and Donato (a.k.a. Armando), every day we think of you and feel your blessings from on high.

Preface

The goal of this book is to provide a rigorous explanation of the principles upon which it is possible to build the organizational transformation from conventional functional hierarchies to *whole system optimization.* In previous books, the authors have laid out the epistemological and methodological framework for this. Here, in "From Silos to Network," we pay specific attention to the *algorithm* that underpins that transformation.

Such a transformation necessitates a radical shift from the obsolete command and control style of leadership to one based on the management of *variation* in every aspect of the life of the organization and the focus on the leverage point of value creation in the system called *constraint.*

We have written this book to substantiate our claim that only an organization designed like a network can withstand the pace of change the business world demands. The readers we had in mind are decision makers that have the need or the desire to truly transform the ability of their organization to compete and who see knowledge, not management fads, as a way to pursue that goal.

This book presents principles that have a universal application, no matter the business sector, and knowledge that has been developed not in academia but through a continuous interplay between theoretical development and in the field validation. We do not include an overview of literature or case histories as examples; we leave that to the Business Schools. With Dr. Deming, we believe that examples teach nothing unless the reader owns the theory. In this book, we aim to unveil the knowledge base supporting the claim that there is a practical way to transition from silo to network.

Ottawa, Canada — Domenico Lepore
Ottawa, Canada — Angela Montgomery
Toronto, Canada — Giovanni Siepe
Baronissi, Italy — Francesco Siepe

Acknowledgements

In our 30 years of quest for a more scientifically rigorous, intellectually evolved and meaningful form of management, we had the great good fortune to meet, and build relationships with, very talented individuals who helped us shape our ideas and the ways with which they could be best communicated.

Our idea of "organizations as Networks of Projects" could only truly come to life if an adequate tool was made available that would engender our sophisticated algorithm. If today we can offer a practical tool for the transformation from Silo to Network, much is owed to the work of Nicola Vastola, founder and leader of Zumbat, and his philosophy of "software with a purpose."

Nicola and his team all through the development of *Ess3ntial* have shown, well beyond their excellent technical skills, an uncanny ability to capture the very essence of the management approach that our algorithm would support and develop a code fit for it; the enlightened, elegant simplicity of the interface is also due to their mastery. We are confident that the enriching and life-enhancing partnership we have developed with the Zumbat team will continue to produce exceptional results and open new possibilities for applications of a systems approach to management.

All through the book we have striven to highlight the foundational role that variation plays in creating sustainability in the management of a network. Our understanding of this subject owes to the work of Dr. Donald J. Wheeler just as much as it does to Dr. Deming's. Don's production is unparalleled for depth and scope. The clarity of his teachings stems from that depth, and we sincerely hope we succeeded in portraying in this book a glimmer of his brilliance.

The findings of this book are the result of a relentless scientific inquiry, the continuous interplay between theoretical speculation and experimental validation. In management, such validation can only happen through fieldwork done with organizations that see the potential for the ideas we propose.

We want to thank in particular:

Chris Maki, founder and leader of MSCP in Sherwood Park, Alberta, Canada, for his vision and leadership in building from the ground up the new Industry of "Heat Management" and providing tangible examples of sustainable management practices in the oil sands.

Laura Calzavara and her ebullient team at BeHubble in Venezia, Italy, for marrying their talent and creativity with the rigor of our management methodology and laying the seeds for a well-deserved and rapid expansion.

Edwin Fjeldtevdt, Markus Skallist, and their team at Favrit in Oslo, Norway, for believing in a "better way" to bring to fruition a truly systemic breakthrough software solution that is already revolutionizing the hospitality industry.

Stefano Righetti at Hyphen, Verona, Italy. Stefano, we feel so blessed to have been part for more than 20 years of that concentration of beauty, inventiveness, and creative madness that you have envisioned and nurtured on the shores of Lake Garda. We are forever indebted to you for making us part of your organization.

A special thanks to our fellow Canadian Dr. Anthony Masys for publishing our chapter on Complexity for Springer, thus creating the possibility to connect with Dr. Hisako Niko and making this book possible.

January 2023 (Shevat 5783)
www.intelligentmanagement.ws
www.ess3ntial.com

About This Book

This book provides the conceptual framework and a comprehensive guide to the principles, methods, and tools for managing organizations in the age of complexity, based on the development and international implementations of the Decalogue management method.

An exponential growth of interconnections and interdependencies defines the scenario that industry, government, healthcare, and organizations at large must now be equipped to tackle; unfortunately, the majority of decision makers lack the relevant knowledge to understand and navigate complexity in every aspect of leadership and management, from strategy to operations. This "New Knowledge" is rarely found within organizations, and it is not taught in Business Schools.

Most organizations still operate within a framework of linear thinking and a traditional hierarchy with separate business functions that inevitably create silos. We present a methodology, "The Decalogue," that portrays a genuinely systemic approach for managing complexity in organizations and value chains through focusing on the management of *variation* (Theory of Profound Knowledge) and a leverage point called *constraint* (Theory of Constraints).

The Decalogue methodology, first published in 1999, lays the foundation for the organizational transformation from silo to system and has been implemented internationally in dozens of organizations of different sizes and in a wide variety of sectors since 1996. This systemic approach leverages the intrinsic process and project-based nature of the work of organizations. Functional hierarchy is replaced by a network-like structure, driven by the goal of the system and governed by a new design of the organization called "Network of Projects."

The transition toward the Network of Projects requires a cognitive shift in the way we view and put to good use human talent and ingenuity as well as a powerful algorithm to orchestrate and synchronize individual competencies. We discuss at length this algorithm, how the Theory of Constraints helps in the cognitive challenges of this shift and also how technology can be used fruitfully to assist with the transformation.

In this book, we provide a robust and sustainable model for organizations to adapt and develop within the complex environments that characterize our contemporary reality. In addition, this model enables organizations to achieve considerably more, operationally and financially, with the finite capacity they have available.

We believe this book will be of interest to leaders and managers of organizations as well as researchers and practitioners in the field of management and organizational design.

Introduction

The first 20 years of this millennium, from the tragedy of the Twin Towers to the global financial crisis, from the pandemic to the folly of the war in Ukraine, brought to the fore the emerging feature of the globalized, tightly and inextricably interconnected world we live in: *complexity* (and our limited ability to deal with it).

When it comes to organizations, the acceleration and pervasiveness of digital technology have made blatantly obvious that the prevailing style of management (command and control) and organizational design (functional hierarchy) are quite inadequate to cope with complexity.

While there is widespread agreement that organizations are complex systems, very few attempts are being made by decision makers to adapt the organizations they lead to this new awareness. Why is this? The majority of Business Schools are demonstrating a structural inability to develop leaders for the age of complexity due to their curricula that remain "ignorant" of complexity and its implications for management. More importantly, the inherent challenges posed by complexity are cognitive; our mind and, more broadly, our "learning apparatus," struggle to grasp the underlying nonlinearity of the phenomena (what we experience) that complexity generates.

Unfortunately, the scientific discoveries made in this field, some of which have deserved the Nobel Prize, still do not bridge the gap between what is valid (there are no truths in science, just validity) and how to use those discoveries to address practical management issues.

Science for Management

Among the scientists that did offer a powerful and profound contribution to the development of the social science of management, the authors believe that Dr. W. Deming and Dr. Eliyahu M. Goldratt stand out of the crowd. Their teachings on quality and flow optimization provide the backbone for what is needed to evolve beyond the current impasse; their work can move management from the quicksand of an ever-present empiricism to a more solid, knowledge-based terrain.

Dr. Deming is unanimously recognized as the founder of the quality movement; his books on management require nothing less than committed study and are widely considered pillars of systemic management. Conversely, Dr. Goldratt presented much of his findings, collectively known as the Theory of Constraints (or TOC), in the form of easy to read business novels, most of which are best sellers (2). While this format gave Goldratt worldwide visibility, somehow, in the process, the epistemological stance on management underpinned by his novels got lost and the Theory of Constraints has become something of a popular cult that is difficult to decipher.

An epistemological stance, instead, is critical in a complex environment; leadership and management need a theory to guide their navigation in the somewhat turbulent waters of the nonlinear, cause-effect relationships that shape the reality of organizations. Just as pilots need instrumentation to fly airplanes across continents, decision makers need a new kind of science to sustain the new economics dictated by complexity.

Since the mid-1990s, we have worked on a cohesive, operational integration of the work of Deming and Goldratt in our Decalogue management methodology. It has formed the basis for the last 15 years of our work. The Decalogue was first presented in 1999 in "Deming and Goldratt: the Decalogue" (3). Our findings, corroborated by a relentless feedback cycle between theoretical development and in the field validations, were first published in a collection of essays called "Sechel: Logic, Language and Tools to Manage any Organization as a Network" (4), then more formally in a chapter called "Managing Complexity in Organizations Through a Systemic Network of Projects" in a volume published by Springer (5) and then expanded in "Quality, Involvement, Flow: The systemic Organization" (6) and later applied to digital transformation in "Moving the Chains" (7).

If nothing else, this book is an attempt to regain clarity on the "why and how" we need a Deming-Goldratt-based approach to sustain innovation and competitiveness in this age of complexity. We call this approach the *Network of Projects*, and its aim is to evolve the science of management put forth by Deming and Goldratt into the realm of organization design.

In essence, a new kind of science for management recognizes that organizations are *systems*, and more specifically, they are *networks of interdependent components aimed at a common goal.* If we arrange these interdependencies so that they are *subordinated* to a strategically chosen *leverage point*, we can maximize the speed of value flow as perceived by the markets to which we sell. We call this leverage point, *the constraint* of the system.

What "subordination" means in this context is that a quality-driven optimization process must be seen as a way to facilitate the work of our chosen leverage point, the constraint, hence increasing the speed of flow.

Quality (and the low-variation processes that ensure it) and speed of flow (dictated by the Constraint) become, then, the cornerstones upon which it is possible to build *the sustainable development of any organization.*

A New Organizational Design for Complexity

In order to engender in the life of an organization the operational awareness of quality and flow, we must move away from the obsolete, functional hierarchy and its cost-driven, budget-focused control mechanism that artificially imposes silos. What must we replace it with? One that is intrinsic to the workings of a system and focuses on the pace (speed) at which the system can generate units of its designated goal, its *Throughput*.

Such control must be based on the inherent feature of every system, the *variation* associated with its components, and, especially yet not exclusively, on how the cumulative variation generated by the system impacts the constraint. At the most fundamental level, management becomes, then, how we protect and control the value flow the constraint can realistically generate. We call it *buffer management.*

The transition from silo to system underpinned by quality and flow needs an *algorithm* that can operationally sustain it as well as a new covenant based on *competencies* with which we *involve* people in the organization and the way we systematically create win-win situations in the business *environment* we are part of (customers, suppliers, shareholders, stakeholders at large). We call this environment, and the network of conversations it entails, the "Network of Projects."

What you should expect from the following chapters is a rigorous explanation of the role that variation and constraint play in the building of a systemic organization based on synchronized projects and how this helps reshape the way we look at the role of Human Resources and Value creation.

Notes

1. See Lepore, D. Cohen, O.: Deming and Goldratt: the Decalogue. North River Press, Great Barrington MA (1999); Lepore, D.: Sechel: Logic, Language and tools to Manage any Organization as a Network. Intelligent Management, Toronto (2010); Lepore, D., Montgomery, A., Siepe, G.: Managing Complexity in Organizations Through a Systemic Network of Projects. In: Masys, A. (ed.) Applications of Systems Thinking and Soft Operations Research in Managing Complexity, pp. 35–69. Springer, Switzerland (2016); Lepore, D., Montgomery, A., Siepe, G.: Quality, Involvement, Flow: The Systemic Organization. CRC Press, NY (2017); Lepore, D. Moving the Chains: An Operational Solution for Embracing Complexity in the Digital Age. New York, Business Expert Press, NY (2019).
2. See: Goldratt, E. M.: The Goal: A Process of Ongoing Improvement. North River Press, Great Barrington, MA (1984); Goldratt, E. M.: It's Not Luck. North River Press, Great Barrington, MA (1994); Goldratt, E. M.: Critical Chain. North River Press, Great Barrington, MA (1997)
3. Lepore, D. Cohen, O.: Deming and Goldratt: the Decalogue. North River Press, Great Barrington, MA (1999)
4. Lepore, D.: Sechel: Logic, Language and tools to Manage any Organization as a Network. Intelligent Management, Toronto (2010)

5. Lepore, D., Montgomery, A., Siepe, G.: Managing Complexity in Organizations Through a Systemic Network of Projects. In Anthony Masys (ed.) Applications of Systems Thinking and Soft Operations Research in Managing Complexity, pp. 35–69. Springer International Publishing, Switzerland (2016)
6. Lepore, D., Montgomery, A., Siepe, G.: Quality, Involvement, Flow: The Systemic Organization. CRC Press, NY (2017)
7. Lepore, D.: Moving the Chains: An Operational Solution for Embracing Complexity in the Digital Age. New York, Business Expert Press, NY (2019)

Contents

About the Authors

Dr. Domenico Lepore graduated in Physics from the University of Salerno (*dottore in fisica*), Italy, in 1988. His research work was published in "Twenty-junction arrays for a Josephson voltage standard at 100 mV level." D. Andreone, V. Lacquaniti, G. Costabile, D. Lepore, R. Monaco, S. Pagano, M. Russo and G. Costabile Eds., World Scientific Pub. Co, Singapore, 1988, p. 1–12.

After several years at the School of Entrepreneurship of the Ministry for Industry in Milan (Formaper), Lepore founded a consulting firm where he and his team developed the Decalogue Management Methodology, a unique synergy of Deming's Theory of Profound Knowledge and Goldratt's Theory of Constraints, successfully implemented internationally in many different sectors. Following a period in the USA as President of a public company, Lepore founded Intelligent Management in Canada to disseminate the Decalogue and develop the "Network of Projects" organizational design. Intelligent Management developed the Ess3ntial platform for managing multi-project environments, drawing on aspects of Systems Science and Network Theory.

Lepore is an author and co-author of several publications, including Lepore, D., Cohen O.: *Deming and Goldratt: The Decalogue.* North River Press, Great Barrington, Mass. (1999) Lepore, D.: *Sechel: logic, language and tools to manage any organization as a network.* Intelligent Management Inc., Toronto (2011); Lepore, D., Montgomery, A., Siepe, G.: *Quality, Involvement, Flow: The Systemic Organization.* CRC Press, New York (2017); Lepore, D.: *Moving the Chains: An Operational Solution for Embracing Complexity in the Digital Age.* New York, Business Expert Press (2019); Lepore, D., Montgomery, A., Siepe, G.: *Managing Complexity in Organizations Through a Systemic Network of Projects.* In Anthony Masys (ed.) *Applications of Systems Thinking and Soft Operations Research in Managing Complexity*, pp. 35–69. Springer International Publishing, Switzerland (2016).

Dr. Angela Montgomery was awarded a Ph.D. in Literature from the University of London in 1994 with a thesis on Literature and Science. She taught at the Universities of London and Salerno and at the European Business School, Milan.

In 1996, together with Dr. Domenico Lepore, she co-founded a consulting firm in Milan, Italy, that developed a systemic management methodology, the Decalogue. She became directly involved in the dissemination of the Decalogue with a particular interest in the Thinking Processes and subsequently co-founded Intelligent Management in Canada.

Montgomery writes regularly about the Decalogue methodology. She has published works on literature such as Beckett and Science: Watt and the Quantum Universe. In: E. S. Shaffer (ed.) *Comparative Criticism, Literature and Science* vol. 13, pp. 171–181. Cambridge University Press, Cambridge (1992) and she has co-authored publications on management including Lepore, D., Montgomery, A., Siepe, G.: *Quality, Involvement, Flow: The Systemic Organization.* CRC Press, New York (2017); Lepore, D., Montgomery, A., Siepe, G.: *Managing Complexity in Organizations Through a Systemic Network of Projects.* In Anthony Masys (ed.) Applications of Systems Thinking and Soft Operations Research in Managing Complexity, pp. 35–69. Springer International Publishing, Switzerland (2016).

Montgomery is the author of a business novel about personal and business transformation called "The Human Constraint" that provides a narrative version of several case histories with the Decalogue method. Downloaded in 43 countries as a digital project, it will be published by Taylor & Francis in 2024.

Dr. Giovanni Siepe graduated in Physics from the University of Salerno, Italy, in 1985 (*dottore in fisica, magna cum laude*) with research work in theoretical physics. He worked for almost two decades as an executive in industries in Italy and Switzerland before joining Domenico Lepore and Angela Montgomery's consulting firm in Milan, Italy, where he led the creation and development of algorithms and software for the management of organizations as complex systems.

In 2007 Siepe became Director of Statistical Studies for Barzel Industries in Canada. He became a co-founder of Intelligent Management in 2010, a consulting firm with the goal of further disseminating the Decalogue Management Methodology. He led and supervised the entire development of the Ess3ntial web-based platform for managing multiple projects at finite capacity.

Siepe is a co-author of several papers and books on managing complexity, including Maci, G., Lepore, D., Pagano, S., Siepe, G.: *Systemic Approach to Management: A case study.* Poster presented at 5th European Conference on Complex Systems, Hebrew University, Givat Ram Campus, Jerusalem, Israel, 14–19 September 2008; Lepore, D., Montgomery, A., Siepe, G.: *Quality, Involvement, Flow: The Systemic Organization.* CRC Press, New York (2017); Lepore, D., Montgomery, A., Siepe, G.: *Managing Complexity in Organizations Through a Systemic Network of Projects.* In Anthony Masys (ed.) Applications of Systems Thinking and Soft Operations Research in Managing Complexity, pp. 35–69. Springer International Publishing, Switzerland (2016).

Dr. Francesco Siepe graduated in Mathematics (magna cum laude) from the University of Salerno in 1993 with a thesis on Caccioppoli sets. Siepe was awarded a Ph.D. in Mathematics from the University of Florence in 1998, with a thesis in the field of Calculus of Variations. He received a research grant at the Faculty of Architecture of the University of Florence, where he carried out research activities on partial differential equations.

He has taught Mathematics and Physics in Italy since 2001 and is adjunct professor at the University of Salerno, teaching and examining Calculus I and II, and Geometry for the Departments of Civil Engineering and Mathematics for Advanced Calculus.

In 2009, Siepe became Project Coordinator and Algorithm Developer for Barzel Industries, Canada.

Since 2010, he has been a consultant for Intelligent Management for whom he developed the entire algorithm for the Ess3ntial platform for managing multi-project environments. This new algorithm represents a significant innovation in the application of the Critical Chain solution from the Theory of Constraints for multi-project environments.

Siepe has produced many international publications on topics relating to the existence and regularity of the solutions of partial differential equations and of functionals of the calculus of variations. Publishers include Istituto Matematico at the University of Trieste, the University of Rome, Manuscripta Mathematica, Zeitschrift für Analysis und ihre Anwendungen, the University of Rosario, Argentina, Journal of Mathematical Analysis and Applications, Mathematical Institute of Charles University, Prague, Czech Republic, and a chapter on Information Systems for Intelligent Management, Canada.

Chapter 1
Shifting Away from Silos

Abstract Most organizations are still based conceptually and operationally on models that are insufficient for complexity. Management remains widely rooted in the idea of command and control, and this is reflected in a traditional, hierarchical/functional organizational design (silo mentality). We present a systemic approach and methodology for management based on the work of physicists Dr. W. Edwards Deming (Theory of Profound Knowledge) and Dr. Eliyahu Goldratt (Theory of Constraints). We introduce an organization design we call the Network of Projects where projects are scheduled at finite capacity from a pool of competencies.

Keywords Complexity · Management · Silos · Deming · Theory of Constraints · Organizational design

It has never been so evident that the major challenge of our times is complexity—a level of interdependencies that the world has never before experienced. However, there is little real understanding in the corporate world, or elsewhere for that matter, of what complexity is, how it originates, and what deep shifts, not just techniques and tactics, are required to operate and innovate in a complex world. Indeed, most organizations are still based conceptually and operationally on models that are insufficient for complexity. Rather, they reflect a mechanistic paradigm that leads to silos, fragmentation and a zero sum game. No matter which techniques forward-looking managers may try to adopt, management remains widely rooted in the idea of command and control, and this is reflected in a traditional, hierarchical/functional organizational design (silo mentality).

1.1 What's Wrong with Silos?

Dividing up an organization as a traditional functional hierarchy that does not take into account the systemic nature of a company and creates a series of problems. Some of these are:

D. Lepore et al., *From Silos to Network: A New Kind of Science for Management*, SpringerBriefs in Complexity, https://doi.org/10.1007/978-3-031-40228-9_1

- Communication is slowed down, even blocked;
- Innovation is stifled by bureaucracy;
- Projects are delayed and go over budget;
- Quality suffers;
- Reaction time to changing market demands is too slow;
- Individuals are delayed in carrying out the tasks for which they are responsible due to a misalignment between responsibility and authority;
- There is no natural way for competencies, both technical and managerial, to develop due to artificial "ceilings";
- Departments carry out their work based on local measurements and therefore local optima, losing sight of the overall goal of the organization;
- Production planning can be completely uncorrelated with sales;
- Cause and effect relationships that exist in organizations are not evident (it takes time for the effects of a cause to propagate through a system);
- There are no means for understanding the implications of local, siloed decisions on the big picture.

1.2 A New Systemic Solution

Ever since the mid 1990s, our work in the field with companies in a wide range of sectors in Europe and North America has had the aim of introducing and continuously developing a *systemic* approach to management that is fit for our age of complexity. This necessarily requires elements of science that are lacking in the day-to-day management of most companies.

We can thank Dr. W. Edwards Deming for enlightening the world to the fact that companies are inherently systems. This understanding enables us to view the organization as an integrated whole instead of a collection of departments or functions with artificial barriers. Consequently, an organization needs to be managed as a whole and in terms of the interdependencies that exist both internally and externally, as demonstrated in the 1950s in Deming's famous sketch of "Production Viewed as a System." The following diagram in Fig. 1.1 takes inspiration from his sketch. Our version shows how a *whole organization* can be viewed as a system.

Deming's definition of a system in *The New Economics* is the most useful and practical:

> A System is a set of interdependent components that work together toward a common goal.

Through our work with the Decalogue methodology, first published in Lepore and Cohen (1999), we have come to recognize over the years that there are just two fundamental forms of activities in any organization: repetitive processes and one-off initiatives, or projects. In light of this realization, we can remain coherent with Dr. Deming's precepts in a way that is rigorous and re-verbalize his definition as:

> An organization is a System; a network of interdependent people and resources working in *processes and projects* to achieve a common goal.

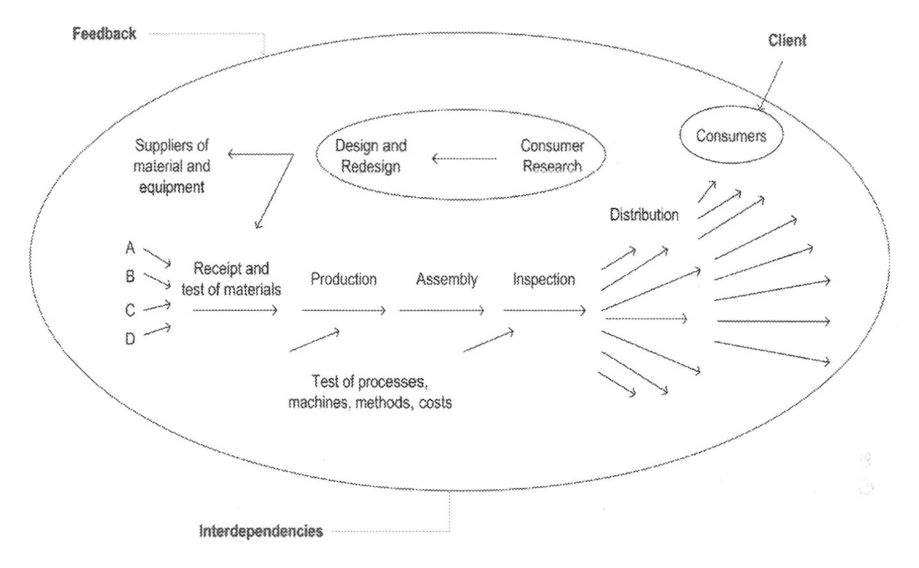

Fig. 1.1 A whole organization viewed as a system

When we take into account network theory, we can provide a more contemporary definition of an organization as: An organization is an *Oriented Network* where the direction of the *orientation* is determined by the goal of the organization (1).

This understanding has implications for all aspects of the organization, from leadership to day-to-day operations.

1.3 How Do We Build a Truly Systemic Organization?

Make everything as simple as possible, but not simpler.

(Albert Einstein)

Everything we do today is immersed in a reality that is complex and we constantly receive a multiplicity of inputs that did not exist 50, or even 20 years ago. For most organizations, the reaction to this is based on something that may have been effective in the past: to try and control complexity by breaking it up into smaller "pieces." The assumption here is that by breaking a big situation up into smaller pieces, more control can be exerted over every piece. This is a very flawed assumption and it leads organizations to design their operations by dividing them up into the silos of departments, functions and too many hierarchical levels. The challenge today, instead, is how to manage unprecedented levels of *interdependencies* that are a characteristic of complexity.

The solution we have developed over the years has focused increasingly on organizational design; it is only at this level that we can truly tackle the ambitious goal of shifting from the prevailing management style built on the traditional functional/hierarchical model and command and control to a paradigm of whole system optimization.

So how do we build a truly systemic organization? There is no a short answer. Nevertheless, we can point out the key elements that we have identified as necessary to achieve this goal.

(1) *Low variation* in processes and projects within a context of statistical predictability (*Quality*);
(2) *Synchronization* of processes and projects through a strategically chosen constraint (Speed of *Flow*);
(3) *Involvement* of all the internal and external stakeholders.

We have described these basic pillars for a truly sustainable *Systemic Organization* at length in our previous publications where we offer a viable, epistemological framework for such an organization. (2).

This solid epistemological framework is what enables us to propose a clear path for building a systemic organization *operationally*. Our focus has been not just to develop a strong theory but to create a coherent methodology with the practical tools to make the transformation happen.

Let's start from the basics.

1.4 Mapping the System

As Dr. Deming used to say, "If you cannot describe what you are doing as a process, you don't know what you are doing".

Each and every part of an organization, be it a person, a machine or even a project or process, has a role in the System. Highlighting these roles, how they interdepend and the flow of activities they are part of is the starting point to make sure, as Dr. Deming recommends, that we know "what we are doing".

Every company has a certain number of activities that it must carry out in order to deliver what it sells, be it a product or a service, to a customer.

Mapping out all the interdependencies/linkages that exist requires us to understand the company's processes and how to link them together. This can be done simply and effectively by using flowcharts to map out every process in the organization, identifying who does what and in what sequence. A flowchart will show the interactions among people in the various phases of the process. It is crucial to know what these interactions are to really understand how the process works and how to improve it.

A flowchart provides a map (or a chain, if you will) of tasks and decisions. It describes the flow of materials, information, and documentation. Ultimately, the main purpose of flowcharting is to show the way the System transforms inputs into outputs. Drawing a flowchart immediately draws attention to the fact that most process chains must cross functional boundaries in order to deliver a product or service to the customer. Flowcharting encourages everybody to describe "how it should be done" and this can only be effective in an environment that is free from finger-pointing and open to change.

This is a first (and crucial) step in understanding that the conventional, functional pyramid (organization chart) that is still today the way so many organizations think of themselves is a self-imposed limitation (an artificial constraint) to the development of a truly Systemic Organization.

Flowcharting processes to show how they should ideally happen is a very revealing activity. It immediately becomes obvious that the existing processes contain unnecessary, often obsolete, loops. These may well have been brought about by the need to create a workaround solution (probably triggered by some emergency) that has then become accepted as common practice. The flowcharting will also show any misalignments that exist between employees' authority and responsibility that prevent them from taking necessary actions autonomously and it will bring to light problems and/or breakdowns in the customer-supplier chain that were previously unclear.

Simply put, flowcharting helps us overcome blockages and barriers that prevent the processes from running smoothly and effectively. Moreover, flowcharts help to identify *Key Quality Characteristics* (KQC), i.e. those aspects of a process that strongly affect the ability of that process to achieve the goal for which it was designed (have a major impact on the performance of the organization as a whole.) We can then use these KQC as points in the process to collect meaningful data to monitor variation. This will inform us:

- whether or not the process is in a state of statistical control before we attempt to improve it;
- whether or not attempts to improve the process are succeeding.

1.5 Deployment Flowcharts

For the purpose of this book we shall consider a particular kind of flowchart called a *deployment flowchart (DFC)*. A DFC shows the interactions among people that take place along the various stages of the process as well as who does what. It is crucial to know what these interactions are to really understand how the process operates and how to improve it.

The operational steps to draw a DFC are:

(1) Define the process boundaries: where will it start and finish?
(2) Identify the key areas of the process (the main competencies required). To help people focus and describe their interactions accurately, we can ask questions such as: What are the inputs to your task that are relevant to the process under discussion? Where does each input come from? What is the expected output? Who is the recipient of the output?

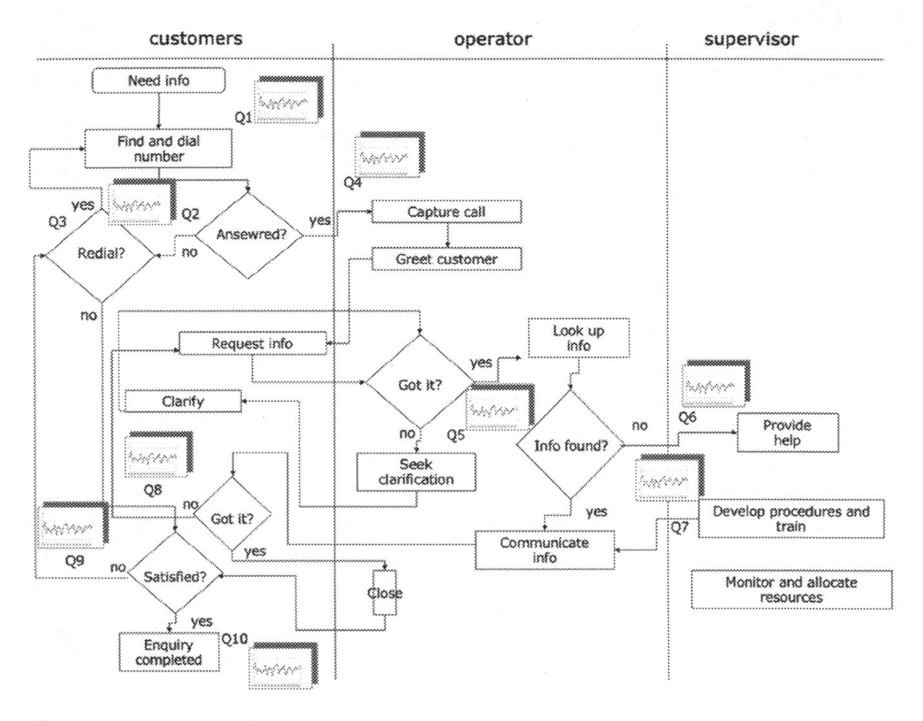

Fig. 1.2 A deployment flowchart

(3) Identify the main roles/competencies required and list them at the top of the flowchart. When the flow moves from one competency to another this is represented by a horizontal line. (Placing the main roles at the center of the page will help reduce oscillation across the flowchart. Otherwise, one should put the main competency involved earlier in the process to the left, and those involved later to the right).

Figure 1.2 is an example of a *deployment flowchart* including KQC. The small charts placed at the KQC points are Process Behaviour Charts. The purpose of these charts is to monitor the variation of processes at those chosen points to learn about the nature of variation affecting them. This knowledge enables us to take rational decisions regarding improvement of the process.

1.6 From Process to Project

As we mentioned earlier, inspired by Deming's definition of an organization as a system, we can state that "An organization is a System; a network of interdependent people and resources working in *processes and projects* to achieve a common goal".

If this is the case, it is important to understand how processes and projects are linked. We can define a *PROCESS* as a set of repetitive activities that follow a given procedure. A *PROJECT* is a special type of process: it is the set of actions (tasks) to be carried out by resources that have the competencies required to satisfy specifications that are often set by a client in an established timeframe and within an agreed-upon budget.

Let's use the following System in Fig. 1.3 inspired by Deming's diagram as an example. We can think of each box in the System as a *center of competency* that represents the *main competencies* necessary to sustain the flow of activities.

For each box we could draw a flowchart to illustrate the interdependencies involved, as in Fig. 1.4.

The question then becomes, how do we create the connection and transition from the repetitive processes in an organization to creating and managing the projects that are necessary to achieve the goal of the organization?

In principle, we could pick a set of resources from the processes, assign them to each task of the project as in Fig. 1.5 and then schedule the project according to the availability of the resources. That may sound like a straightforward solution. Unfortunately, this "solution" would be based on "linear thinking" (where reaction is proportional to the action).

While it may work sometimes, most of the time it would not. We face a problem of complexity, where nonlinear interactions play an important role.

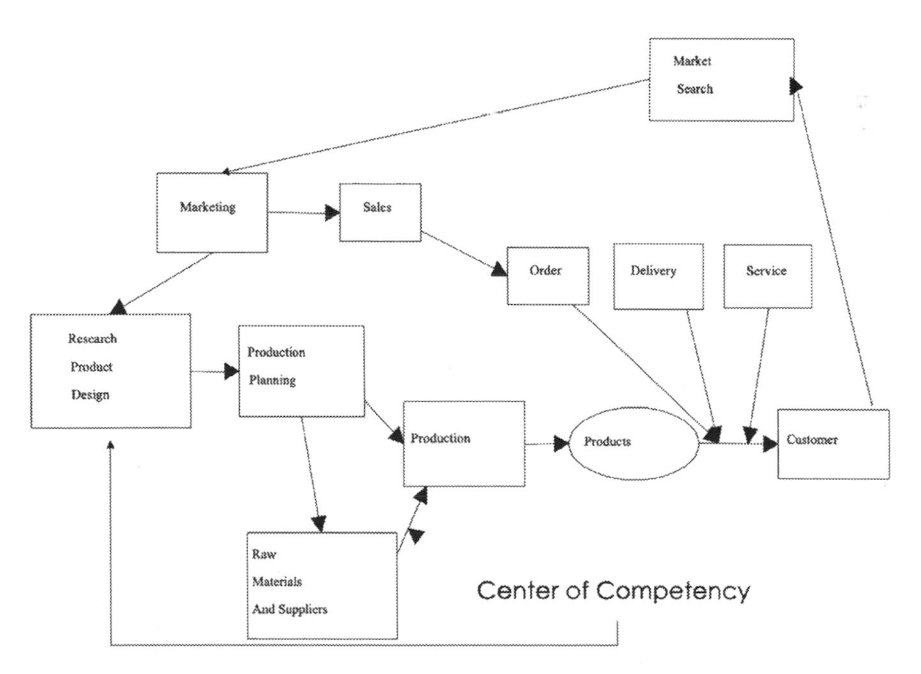

Fig. 1.3 Centers of competencies

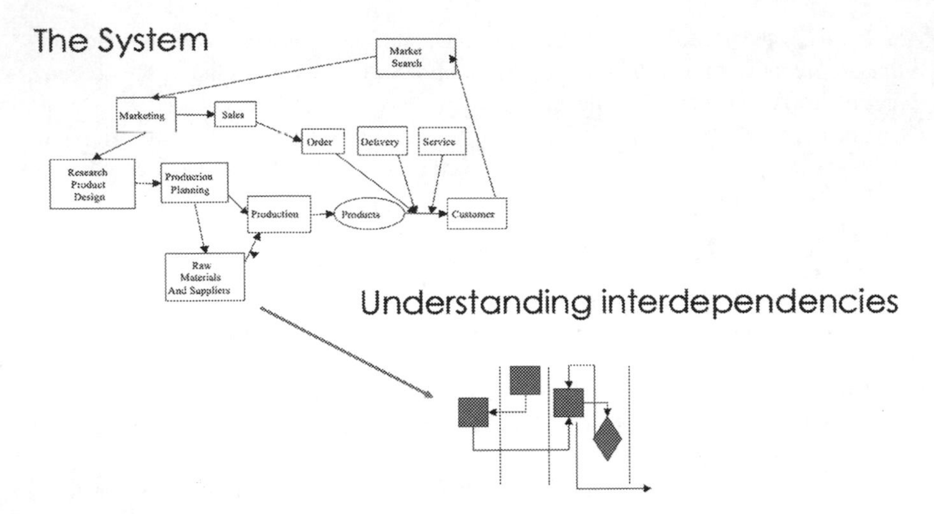

Fig. 1.4 Understanding interdependencies

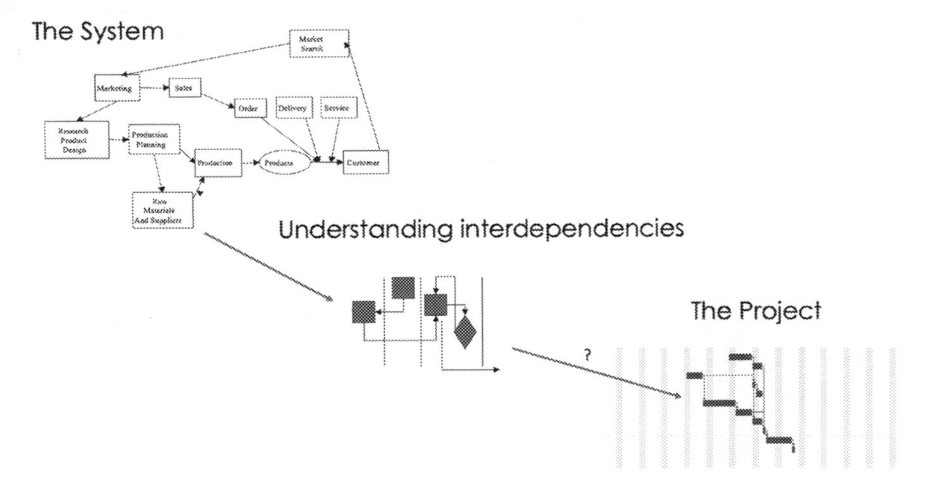

Fig. 1.5 From process to project

1.7 Projects and Complexity

Organizations are *complex systems* and the people who work in them live and evolve in a *complex environment*.

Complexity can neither be simplified nor eliminated. In an organization, we have to learn how to embrace, live with, and manage complexity. We need an approach guided by *non-linear* thinking (where a reaction is not proportional to the action).

This poses many challenges. Generally speaking, there is little or no awareness of complexity and its implications in business and management. It is certainly lacking in the traditional world of Project Management. Consequently, embracing complexity becomes a problem of knowledge but also a problem of handling the cognitive and emotional issues connected with embracing a way of managing fit for complexity. This is the challenge represented by the transition from Processes to Projects.

Dr. Eliyahu Goldratt, who developed the Theory of Constraints, addressed the problem of Project Management from a paradigmatic perspective and developed a breakthrough, *systemic* solution he called Critical Chain. We have taken the Critical Chain solution to the level of organizational design by challenging the assumption that projects are scheduled based on available resources. In our work, we look at available *competencies.*

1.8 Projects and Competencies

At the most basic level, the question the Theory of Constraints (TOC) addresses is: what can we do with what we have? How can our resources generate the most sustainable value? When we talk about individuals, the question becomes: How can we maximize the value people bring to the System?

We feel very strongly that, along with their ingenuity and passion, the highest value people bring to their organizations is found in their *Competencies.* These competencies are both the ones they currently possess and those they can acquire. (Later in this book we will consider what Human Resource management should focus on: identify, nurture and evolve the competencies the organization needs in order to sustain competitiveness over time as this is how they impact the bottom line).

Competencies are not limited to what people write in their CVs, what they have studied, their work experience, etc. For instance, people in Accounting could easily carry out relatively simple tasks that are completely unrelated to their day-to-day activities, such as supporting Sales in extracting and ordering a list of customers from a database according to a defined criterion. It is important to understand that this kind of systemic collaboration would be impossible in a siloed organization. It would be perceived as a "distraction" of resources from the business function to which they "belong". In a "functional" organization the only optimum possible is local, NOT global.

The above database task example could easily represent a task in a project where competencies are associated with *individuals;* this is what facilitates *Involvement* as it allows people to exercise their competencies in a much broader context than the artificial confines of a business function and it is the catalyst for the implementation of a truly *Systemic Organization.*

The diagram in Fig. 1.6 illustrates the way competencies are associated with every task to be scheduled in a project.

This is the first step towards overcoming the inherent limitations of a conventional hierarchy: Competencies are associated with *individuals* instead of being fished from a business function to which they "belong". An organization will be populated with

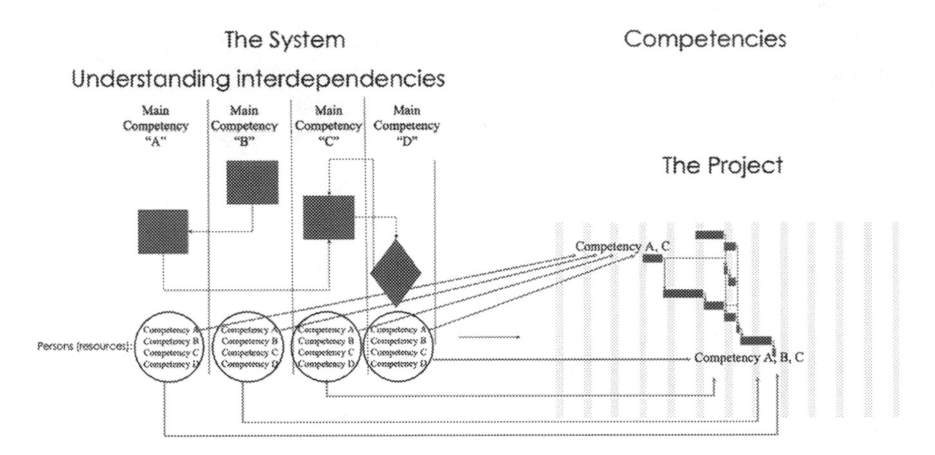

Fig. 1.6 Competencies and tasks

resources that have various competencies. Once we map the competencies of each resource, we can also identify the different levels of the competence the resource possesses, e.g. high level, medium level or low level.

The levels of competencies open up new opportunities for flexibility in the way we schedule the tasks of a project. A resource traditionally *assigned* to the "Center of Competency A" will have a high level of "Competency A", but may also possess a low level of "Competency B". If a task in a project does not require the highest level of Competency B, we can use the resource with a lower level of Competency B, even if that resource *belongs* to the center of Competency A. If we require Competency B, we do not always need to look for the resource in the Center of Competency B. We can look for the right level of Competency B that is available *anywhere in the System* at the required time.

This is key in a multi-project environment as it *frees up capacity that would otherwise not be used* and it facilitates the *multiplication of projects* that can be managed simultaneously.

1.9 The Role of the Constraint

As we stated, we have identified the two fundamental set of activities in any organization as:

(1) Repetitive Processes
(2) New initiatives (one-off projects).

For repetitive processes, it is crucial to describe the flow of activities and monitor variation at some key points in order to manage the System. Low variation in processes is what enables a framework of statistical predictability and it is our main guideline. In Deming's words: "If I could reduce my message to management to just a few words, I'd say it all has to do with reducing variation" (3).

For projects, instead, we focus on the *synchronization* of the different steps of a process through the strategic choice of a physical constraint to which the whole System has to *subordinate*.

Before we consider the constraint of a project, let's define the "Constraint" of a system.

The goal that a System pursues will vary depending on the type of organization, whether it is, for example, a for-profit company, a Research Centre, a school or a hospital, this goal may be cash, new ideas, new solutions, larger market share, healthy patients, knowledgeable students, etc. The *Constraint* is the element that *dictates the pace* at which the System generates *units of the goal.*

There is a precise name for this pace in the Theory of Constraints: *Throughput.* Throughput is a *time derivative*, i.e. units of the goal per unit of time.

The Constraint *is not* a bottleneck as in Fig. 1.7; it is a *leverage point* in the network (see Fig. 1.8). It is the point we strategically choose to generate the highest *value* for the System. It is an important and delicate choice because we have to subordinate the entire System to the Constraint so that it can always work effectively. This entails creating protection capacity in the System. This is because any minute lost on the constraint is a minute of lost throughput that can never be recuperated.

How do we manage a Constraint?

Dr. Goldratt (who began the development of the *Theory of Constraints* (TOC) in the second part of the Seventies) defined a process named *The Five Focusing Steps* to manage the Constraint.

1. *Identify* the constraint—choose it strategically.
2. *Exploit* (leverage) the constraint—make it work as much and as flawlessly as possible on the optimal product mix—the one with the highest yield.
3. *Subordinate* to the constraint—build the whole network of interdependencies in a way that has the statistical capability to support the optimal functioning of the constraint.
4. *Elevate* the constraint—increase its capacity.
5. *If* the constraint has moved (meaning: if its increase in capacity shifts the existing constraint to a new one) then *go back to step 1.*

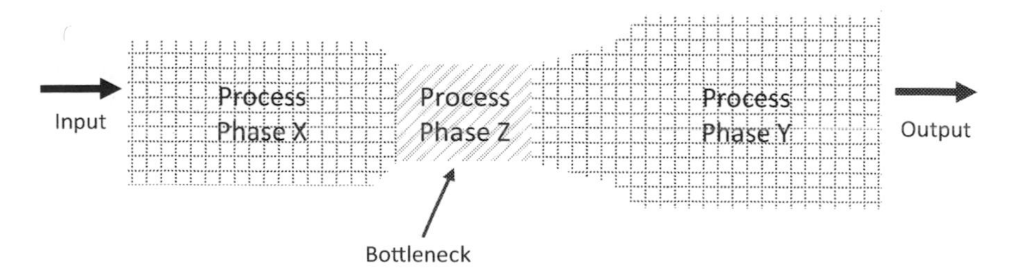

Fig. 1.7 A bottleneck

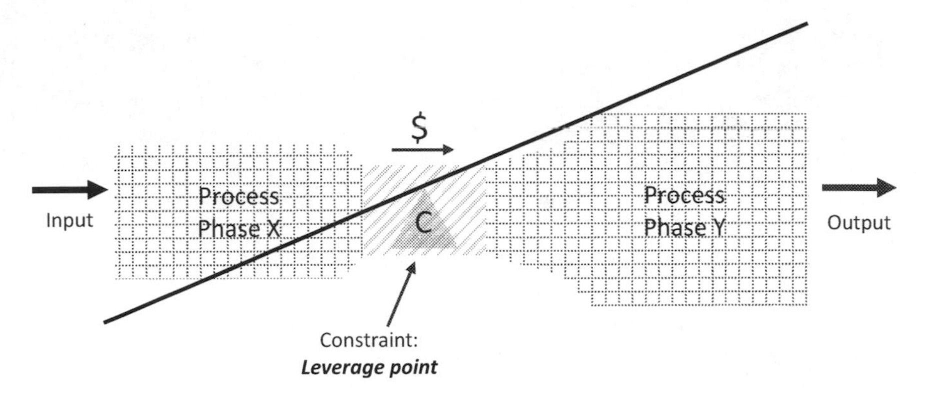

Fig. 1.8 A constraint

Elevating the Constraint (increase its capacity) implies a broader work on the System in order to keep the Constraint in its strategically chosen position. Basically, before we decide to elevate the strategically chosen constraint, we must increase protection capacity where necessary in the System so that the constraint *does not shift*.

It is important to understand that if the constraint shifts, then the process of subordination has to be carried out all over again. This, almost unavoidably, creates disruption in the System. It is for this reason that, once we have *strategically chosen* the Constraint, we really want to keep it where it is. (See: Lepore (2010), p. xvii).

1.10 The Constraint of a Project

When we talk about projects, the concept of constraint is a little different.

A project is a set of actions required to satisfy *specifications* (often set by a client), in an *established timeframe* and within a *predefined budget.*

More precisely:

> A project draws different competencies together in order to accomplish a specified goal, within an agreed upon timeframe and a designated cash outlay.

In the Theory of Constraints, the constraint of a project is its *Critical Chain,* i.e. *the longest sequence of dependent events that takes into consideration the availability of resources* (and in our approach, *competencies).* This chain also defines the lead time for the project's execution, i.e. the realistic length of the project.

This definition is completely coherent with the general definition of constraint because the shorter the critical chain, the faster the project generates value for the organization (Throughput).

1.11 The Buffer

We have defined the Constraint of the System as the element that *determines the pace* at which the System generates units of the goal and any minute lost on the constraint translates into loss of Throughput that will never be regained. Accordingly, the constraint must be protected and the process to protect it is called *Buffer Management*.

The *Buffer* has a dual role: *protection* and *control*.

In a system that is built around and subordinated to the constraint, the buffer protects the constraint from variation that may cause disruption and the conventional way of sizing the buffer is time. In essence, to make sure that the constraint never "starves", a buffer of time is placed in front of it so that breakdowns in the system will not affect the working of the constraint (see Fig. 1.9).

At the same time, we exert control on the System as a whole by statistically monitoring the oscillation of the buffer (more about this later) as in Fig. 1.10.

This line of reasoning is easily grasped for a typical production flow: basically, it translates into a "complete batch of material in front of the constraint, ready to be processed" a "buffer time ahead" of the moment when the constraint should start working on it.

This concept requires some further considerations when it comes to projects. As we said earlier, the constraint of a project is its critical chain. Accordingly, how do we protect the critical chain and at the same time exert effective control all through the execution of the project?

A project is a set of actions required to satisfy specifications (often set by a client), in an established timeframe and within a predefined budget. Project managers inevitably experience the dilemma between adding protection time to the tasks on

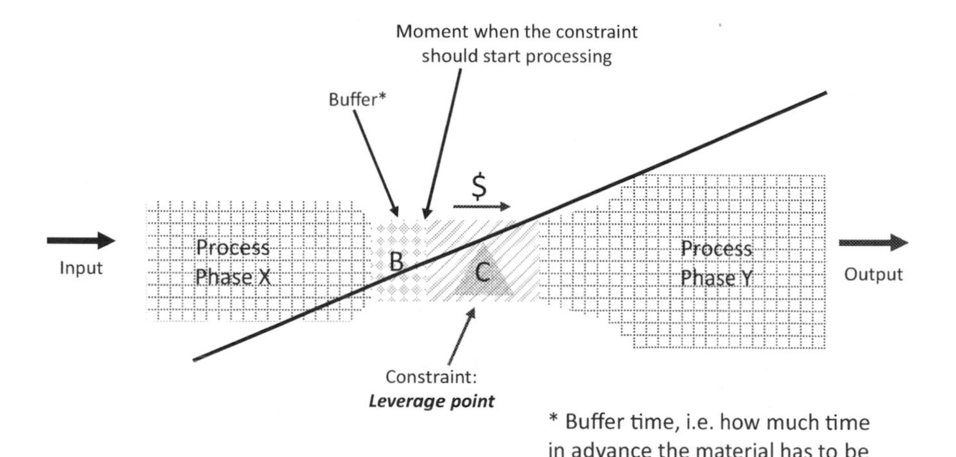

Fig. 1.9 The buffer

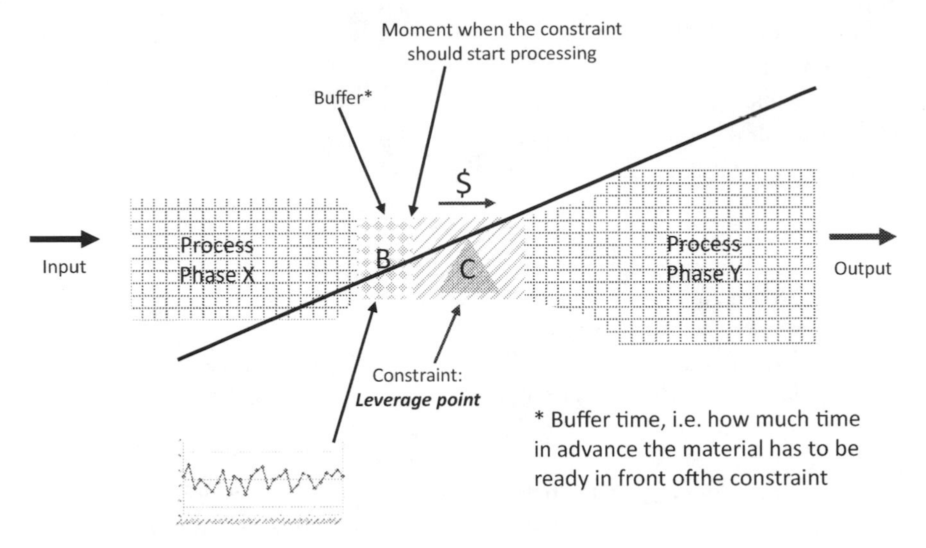

Fig. 1.10 The buffer and statistical process control

the critical chain (to avoid being late) versus NOT adding this protection time (to declare a shorter lead time for completion.)

The main assumption upon which this dilemma rests is that "all the tasks need protection" and, over time, we have come to call this a *cognitive constraint*, something that can be rationally disproved but persistently lingers in people's behaviour.

Dr. Goldratt developed the solution (in Theory of Constraints jargon this kind of solution is called an *injection*) to this dilemma and named it *Project Buffer* (see Appendix A).

In practice, protection only needs to be added to the sequence of tasks that determines the length of the project (the critical chain, i.e. the constraint of the project). Instead of protecting every single task in the critical chain, the protection is cumulated at the end of it in what is called the Project Buffer (see Fig. 1.11).

One can easily dispel any cognitive dissonance between what is presently common practice (wrong way) and what, instead, should be done (right way) by resorting to some well consolidated, universally accepted knowledge.

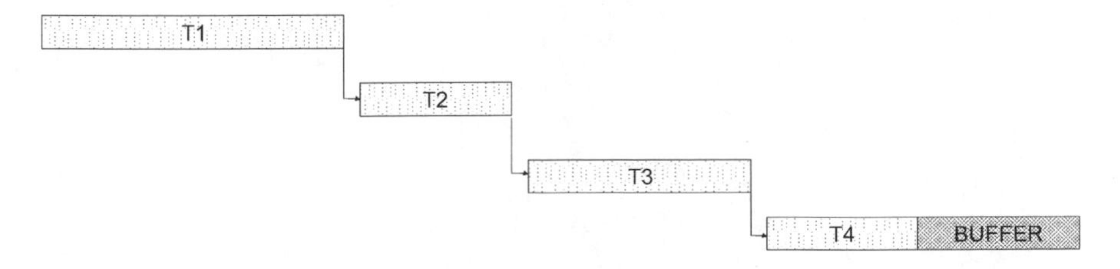

Fig. 1.11 The project buffer

If we have a chain of tasks and we add protection to each single task, what is the difference from adding the protection to the global chain? (See Fig. 1.12).

The answer can be found in Pythagoras' Theorem, normally taught to seventh graders (and some foundational elements of Probability taught in undergraduate courses).

The *cumulated variation* that affects a set of dependent tasks (*Standard Deviation*) decreases at the same rate of the square root of the number of tasks N as in in Fig. 1.13.

Instead of needing "$4 + 3 = 7$", we need "$\sqrt{(4^2 + 3^2)} = 5$".

As a general notation, we need a protection that is $\frac{\sum \sigma_i}{\sqrt{N}}$.

This is how we can effectively protect the completion of a project without unnecessarily inflating its length. Knowledge is the only cure for superstition.

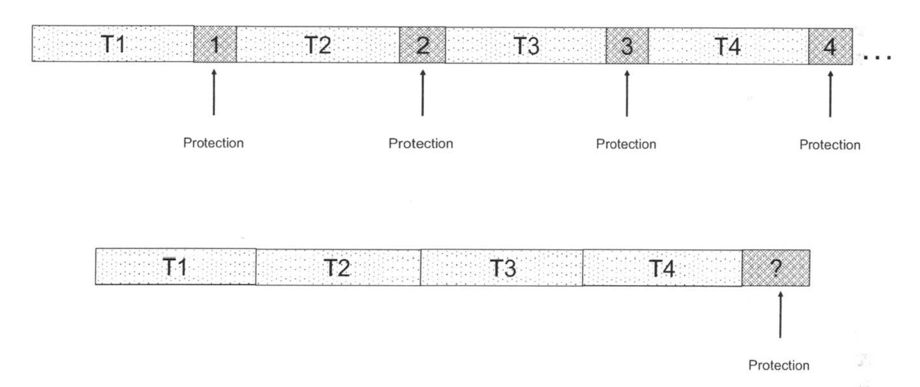

Fig. 1.12 Adding protection

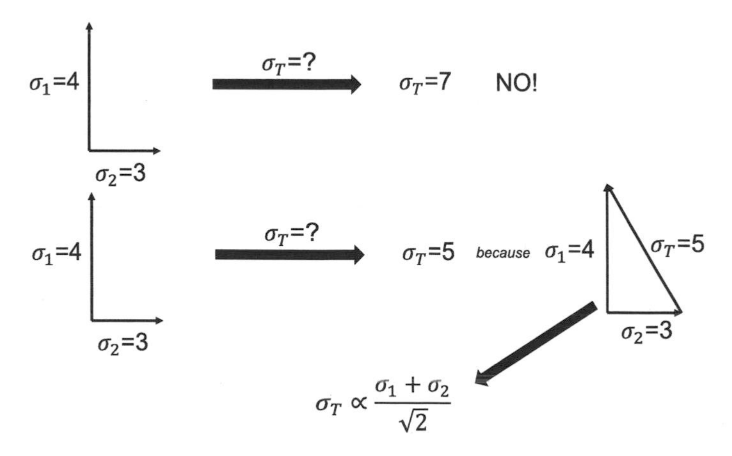

Fig. 1.13 Cumulated variation

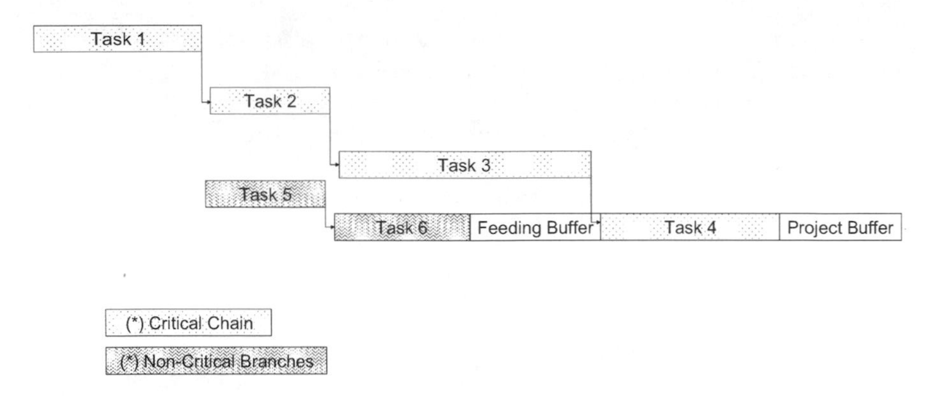

Fig. 1.14 Feeding buffer

1.12 Feeding Buffers

What should we do about protecting other branches in the project that feed the critical chain?

The project buffer is not the only buffer we use to protect the project from the disruption brought by excessive variation. In a project network there are also non-critical branches that can affect the critical chain if their completion is delayed. In this case, the dilemma faced by the Project Manager is between an Early Start of the non-critical branches (to ensure they do not impact the Critical Chain) versus a Late Start (to avoid too much "work in progress").

Here, the cognitive constraint that underpins this dilemma is represented by the assumption that non-critical branches (Feeders) impact the finishing time of the project.

The solution (or *injection*) to this conflict is the *Feeding Buffer* as in Fig. 1.14: non-critical branches are started a buffer-time ahead; in other words we protect the on-time completion of Feeders with an appropriately sized buffer (see Appendix B).

1.13 Managing Projects

Project Management is widely considered a "technique" that is known to Project Management specialists, none of whom are in the "C-Suite". An impressive plethora of project management tools have been developed to aid this technique. These tools are often accompanied by an array of colourful bells and whistles. Unfortunately, virtually none of them addresses the real issue at stake: in an organization that sees itself as a System (as opposed to a functional hierarchy) project management is what Management of the organization is all about. It is the kind of management that engenders the new covenant necessary to embrace Complexity, where organizations are recognized for what they intrinsically are, *Oriented Networks of Projects* (and processes, for that matter) where what determines the direction of the orientation is the goal of the organization.

A new kind of science for management is necessary and this science requires a different set of behaviours in project management (and beyond), as well as an appropriate mindset to accompany them.

Two behaviours that are typical in traditional Project Management but quite inappropriate for a systemic approach to Project Management have to be tackled head on, right from the start of any transformation effort.

The first is connected with the idea of milestones and Dr. Goldratt named this behaviour *student syndrome* (something that many people are very familiar with).

The empirical chart in Fig. 1.15 stresses the fact that the amount of effort expended in the execution of a task is never even, and it follows the trend illustrated.

Inserting milestones throughout the project triggers precisely this kind of behaviour. What is needed, instead, is a constant amount of effort during the execution of each and every task.

The second, probably even more damaging behaviour that is typical in traditional Project Management is *multitasking* as this is very often mistaken for heightened efficiency. Multitasking is instead, as many studies suggest, a major source of a diminished ability of the prefrontal cortex when it comes to the ability to focus. As shown in Fig. 1.16, we can achieve a unit of the goal (Throughput) faster by working on tasks *sequentially* rather than working in multitasking mode.

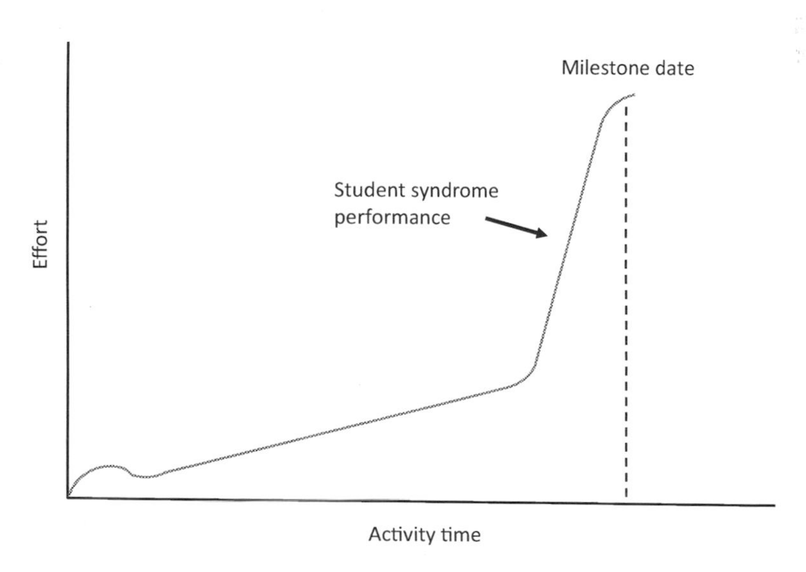

Fig. 1.15 Student syndrome

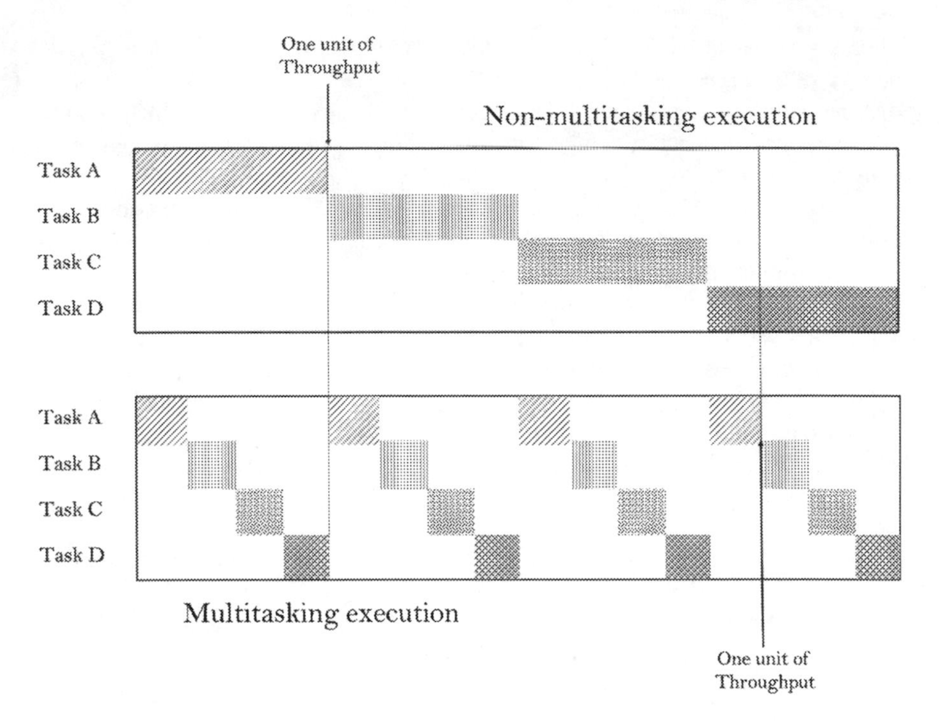

Fig. 1.16 Multitasking

Working in multitasking mode has several negative effects and two major ones are:

(1) It takes much longer to generate the first unit of the goal;
(2) It makes the perception of the duration of a task much longer than it is in reality (this is one of the reasons why during the planning phase of a project people tend to overestimate the time duration of tasks).

1.14 Systemic Project Management and the Organization as a Network of Projects

Why do we claim that a *project is a system?*

One more time (*repetita iuvant):* A project is a set of actions required to satisfy specifications (often set by a client), in an established timeframe and within a predefined budget.

Hence, a project is a set of interdependent tasks that must be carried out within a precise framework of time, budget and specifications. Similarly, a System is a framework designed so that a set of interdependent processes can achieve a common goal.

Therefore, we can safely say that a project is a System and it should be considered as such. Accordingly, how can we then move from the conceptual to the practical? How should we manage this system, operationally?

> The new, systemic management paradigm that we advocate calls for looking at an organization as a "project environment", shaped as a *Network of Synchronized Projects* where people's competencies are scheduled (deployed) based on their real availability (in TOC jargon, at *finite capacity*).

The meaning of *finite capacity* is that there can be no contention in the deployment of resources (as predicated by the Critical Chain algorithm) and when it comes to management, a resource is a supplier of competencies. Every resource has at least one competency, and every competency has a level, as in Fig. 1.17.

In this approach, critical to achieve a truly Systemic Organization, we move *from scheduling resources to scheduling competencies.*

For every resource we establish all the relevant competencies, their level of proficiency, and we then schedule these competencies at finite capacity, using the Critical Chain algorithm. The resources and their competencies are pooled and, when available, can be deployed in every company project, hence overcoming the logic of functional silos.

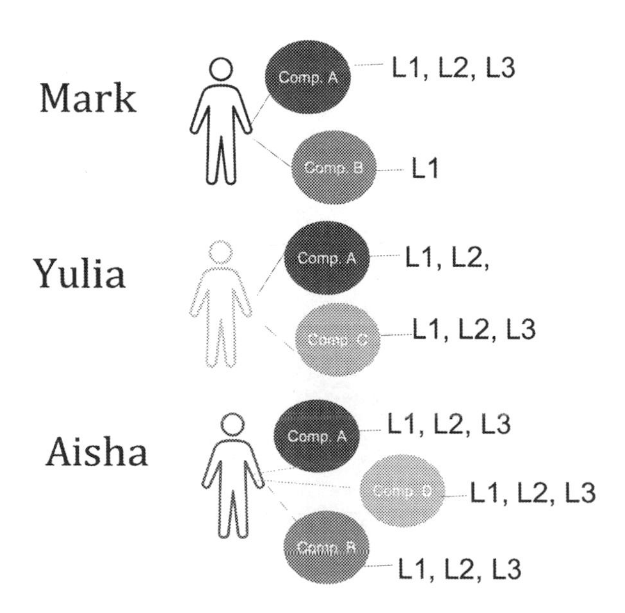

Fig. 1.17 Resources and competencies

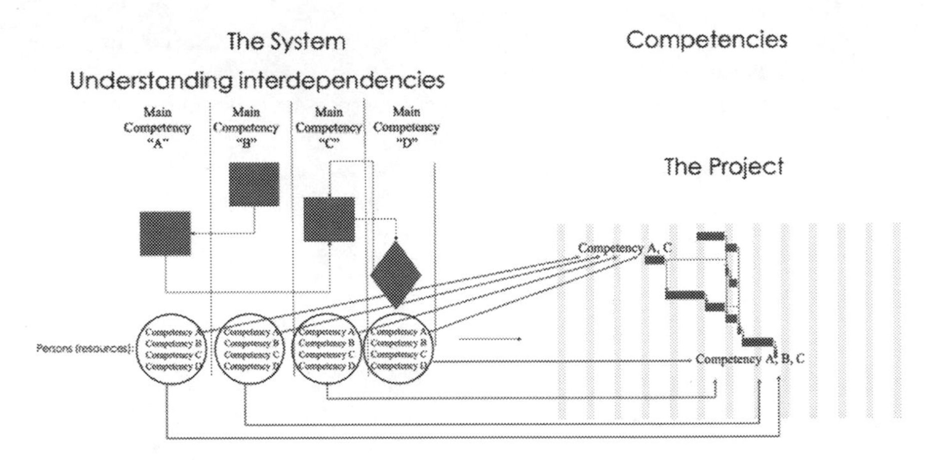

Fig. 1.18 Competencies must be picked from the system

At this point we are ready for the transition from processes to projects in a multi-project environment.

Fig. 1.18 is a graphical representation of the way competencies must be picked from the System to feed a project.

1.15 Planning and Execution

Planning and execution of a project is a ten-step process:

- *Establish* the delivery date.
- *Decide* the size of the Project Buffer (as a percentage of the Critical Chain).
- *Make* a list of the competencies required for the project as well as a list of internal and external resources realistically assignable to the project.
- *List* competencies and levels for each resource.
- *Consider* the calendar of each resource (for availability).
- *Define* Project Tasks and *assign* Duration and Competencies (not resources) to each task while engendering a protocol of no-multitasking.
- *Arrange* the Tasks into a sequence (network) highlighting the logical dependencies among them.
- *Run* the algorithm to determine the Critical Chain. If for any reason we are not satisfied with the schedule, edit the network to make changes.
- *Launch* the project.
- *Update* the progress status of critical ongoing tasks (task update).

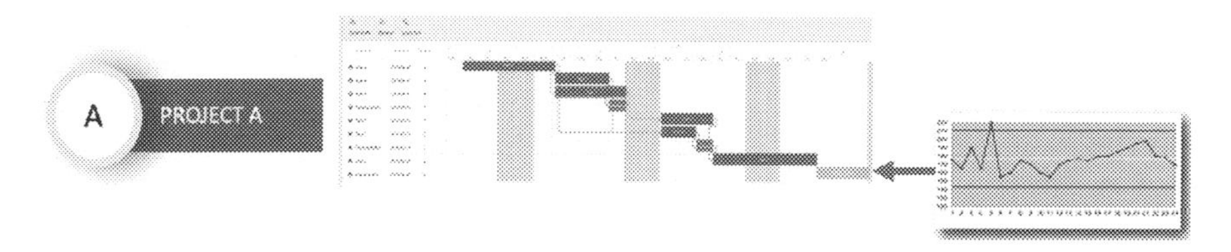

Fig. 1.19 Process behaviour chart to monitor oscillation of buffer consumption

The end result of this process is the GAANT of the project. In the example below, red represents the Critical Chain, at the end of which we place the project buffer. The small chart shown to the side of the project buffer in Fig. 1.19 is a Process Behaviour Chart to monitor the oscillation of the buffer consumption.

When we have more than one project, we will have different Critical Chains for each of the projects as in Fig. 1.20.

In a multi-project environment, we can see the whole network of projects in one chart.

Here is an example in Fig. 1.21 of a global view of the organization as a Network of Synchronized Projects.

Fig. 1.20 Different critical chains for each project

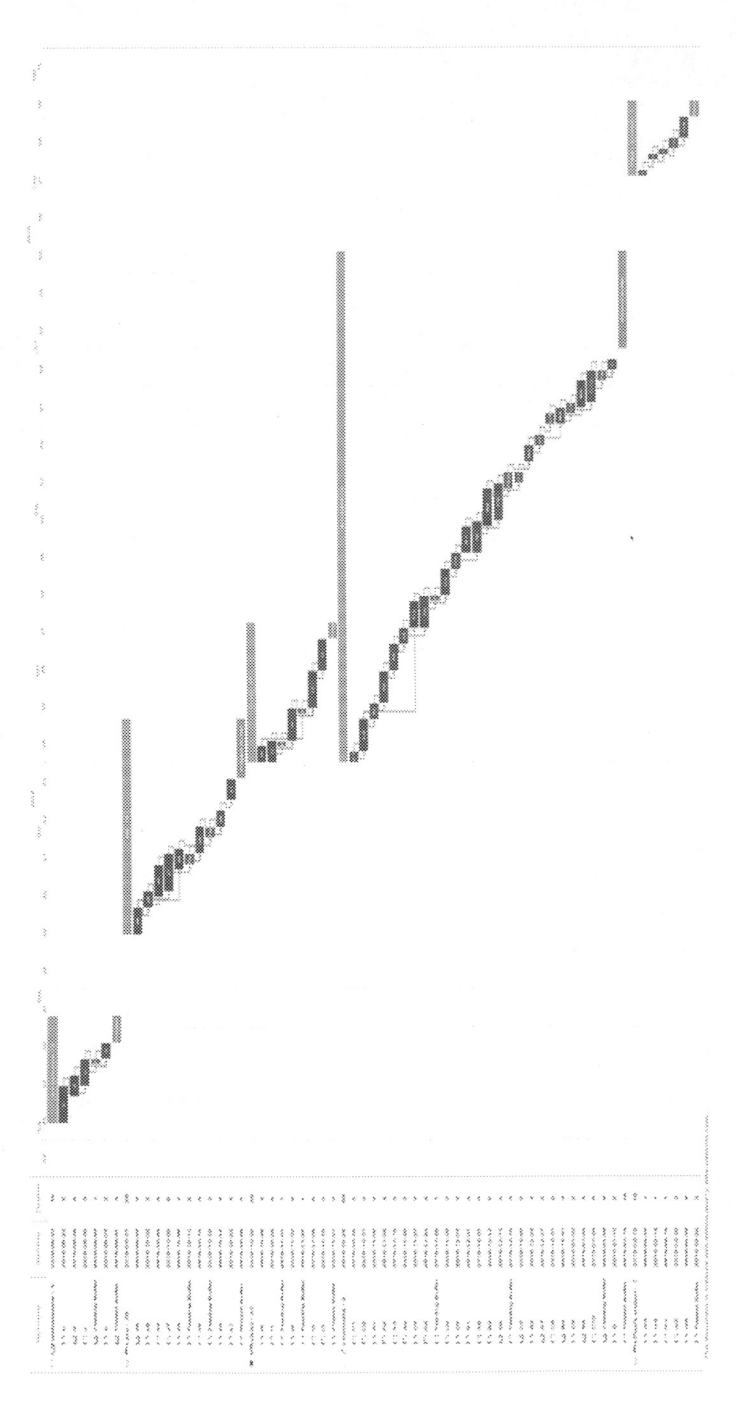

Fig. 1.21 Global view of the organization as a network of synchronized projects

1.16 A New Critical Chain Algorithm–Ess3ntial

As we developed our work with organizations to introduce a systemic organization design that we came to call the Network of Projects, we realized that we would need an appropriate, practical way to support the management of a truly multi-project environment. While there are several software products available that are based on Critical Chain, they would not be suitable for our solution. We needed a truly multi-project solution that would allow companies to manage their whole organization as a network of projects using a pool of competencies. It was a very different way of thinking about projects, what they are for and how they are managed. This led us to the conclusion that we would have to develop our own algorithm and software that would be fit for this purpose. Starting from the math, we developed the algorithm and built the software we have named Ess3ntial.

Ess3ntial is a true multi-project platform that supports the management of an organization designed as a network of projects by using the Critical Chain algorithm.

The ten-step process for managing and executing a project mentioned in the previous section describes how the software works and later in the book we will see a practical example.

First, some further considerations on the meaning of Constraint in a multi-project environment are in order.

The Critical Chain algorithm was developed by Dr. Goldratt and published in an eponymous, highly successful, novel in 1997. The algorithm is based, as are all the algorithms of TOC, on the idea of *constraint*, what we could poetically (and biblically) describe as an inherent limitation towards a goal.

As far as projects are concerned, the constraint is the *Critical Chain* (hence the name of the algorithm), meaning *the longest sequence of dependent events that takes into consideration the availability of resources.* When the algorithm was first developed there was no mention of competencies, but only of how to schedule resources at finite capacity.

The multi-project version of the original algorithm relies on the so-called "pacing resource" (or pacing pool of resources), i.e. the resource that determines the length of all the individual Critical Chains, in order to maximize the overall capacity of the system. It is the designated *scarcest resource* at our disposal that dictates the pace of Throughput generation organization-wide.

The multi-project version never really worked, despite the myriad of workarounds, simply because it could not. Why is that? In a truly multi-project environment, the demand for projects in a given span of time (a week, a month, a year, etc.) is not as *regularly recurring* as the demand for products in a manufacturing environment can be.

Projects are stacked in the system one after the other at any time during the period in consideration, and each project may require the use of an entirely different amount and type of competencies with different delivery dates. This would make identifying a "pacing resource" or a "pacing competency" virtually impossible.

What we came to realize was that in a real multi-project environment, the length of the Critical Chains is determined by *the total amount of available time of the competencies in the System in a given period of time.*

In our quest for a new, truly systemic organizational design, we had to surrender to the idea that a new algorithm was needed. This is exactly what we developed, as we will attempt to explain.

Notes

1. We discuss the nature of networks in the context of their relationship to organizations, variation and the organizational design *The Network of Projects* in our chapter 'Making the Change Operational: The Network of Projects' in Lepore et al. (2017).
2. See Lepore and Cohen (1999), Lepore (2010, 2019) and Lepore et al. (2016, 2017).
3. Edwards Deming (1990).

References

W. Edwards Deming, (Ref: Neave, H.R.: The Deming Dimension) (SPC Press Inc., Knoxville, TN., 1990), p. 57

D. Lepore, O. Cohen, *Deming and Goldratt: The Decalogue* (North River Press, Great Barrington, MA, 1999)

D. Lepore, *Sechel: Logic, Language and Tools to Manage Any Organization as a Network* (Intelligent Management Inc., Toronto, 2010)

D. Lepore, A. Montgomery, G. Siepe, *Quality, Involvement, Flow: The Systemic Organization* (CRC Press, New York, 2017)

D. Lepore, A. Montgomery, G. Siepe, Managing Complexity in Organizations Through a Systemic Network of Projects, in *Applications of Systems Thinking and Soft Operations Research in Managing Complexity*. ed. by A. Masys (Springer International Publishing, Switzerland, 2016), pp. 35–69

D. Lepore, *Moving the Chains: An Operational Solution for Embracing Complexity in the Digital Age* (Business Expert Press, New York, 2019)

Chapter 2
Variation, Buffers and How We Think

Abstract To design organizations as Systems we must understand what kind of variation our processes are affected by as all managerial decisions should be informed by this knowledge. This is important as it is cognitively challenging for decision-makers to deal with probabilistic uncertainty. We look at how to distinguish special causes of variation from common causes of variation and the reasoning that underpins Process Behaviour Charts. We introduce a new method for monitoring the consumption of project buffers as a more rigorous alternative to the three-zone method in the Theory of Constraints. We outline the Network of Projects as an organizational design that involves an oriented network and the effect of complexity on a Network of Projects organization.

Keywords Critical chain · Variation · Process behaviour charts · Buffer management · Oriented network

> Rational behavior requires theory. Reactive behavior requires only reflex action.
>
> (W. Edwards Deming)

> Nothing is permanent in this wicked world, not even our troubles
>
> Charlie Chaplin

When Dr. Goldratt's business novel 'Critical Chain' (1) was published it raised an almost instant planetary interest. This did not come as a surprise because the Project Management world had been for a very long time a fairly stagnant pond that was uninspiring and largely disappointing.

Thousands of project managers all over the world flocked to Dr. Goldratt's conferences and a variety of well-meaning, Critical Chain inspired software products were developed to support Dr. Goldratt's findings. (Some of the authors of this book were involved first hand in the frenzy of those years and cherish receiving the first galley ever produced of the 'Critical Chain' manuscript, given to us by its author).

D. Lepore et al., *From Silos to Network: A New Kind of Science for Management*, SpringerBriefs in Complexity, https://doi.org/10.1007/978-3-031-40228-9_2

While some large corporations fruitfully applied (and are applying) the tenets of finite capacity scheduling, by the beginning of the century the Project Management pond had gone back to its previous stagnation and the rampant lunacy of conventional project management (and related tools) was back in full swing.

In those days of reflecting on why this was happening, one of us had a very heartfelt (and somewhat heart-wrenching) conversation with Dr. Goldratt about the Theory of Constraints. His precise words were: "The TOC community does not understand variation."

May we say many years later, they still do not understand variation and the attempt of many consultants to bundle together TOC, Lean and Six-Sigma is a testament to the all too human ability to generate unnecessary confusion (intellectual and otherwise).

In a chapter called "Managing Complexity in Organizations Through a Systemic Network of Projects" written for Springer a few years ago, (2) the authors of this book highlighted why understanding variation and constraints is key to managing complex organizational systems and the Network of Projects we propose is the offspring of that understanding.

The next few pages are dedicated to anyone who is keen to make a little effort in return for some clarity on this issue.

2.1 What We Need to Know About Variation

In the real world no recurrent events repeat themselves in an identical manner. The outcome of every measurable (repetitive) process will invariably yield different results, measurement after measurement (provided that the measuring tool is sufficiently sensitive). This applies to journey time from the same two locations, daily production output, daily oscillation in body weight, etc. Variation exists.

In his ground breaking book 'Economic Control of Quality of Manufactured Product' (3), Dr. Walter Shewhart laid the foundation for understanding Variation. One of his main findings is that there are *two kinds of variation* (our wording).

1. Variation due to causes intrinsic to the process/system that generates them. They are inherent in the way the process is built and they will always be present. They are called *common causes of variation (or chance causes).*
2. Variation due to external, assignable, causes that are not an inherent part of the way the process is built. They are called *special causes of variation (or assignable causes).*

It is essential to recognize and understand the difference.

Common causes generate over time a pattern that is statistically predictable; we can somehow tell how the process will behave in the future by looking at how it has behaved in the past.

Special causes, on the contrary, do not generate any pattern; if there were any pattern it would allow us to make a prediction and instead the past behaviour of the process will tell us nothing about its future.

If we want to design organizations as Systems it becomes then imperative to understand what kind of variation our processes are affected by as our managerial decisions should be informed by it. Accordingly, how can we distinguish special causes of variation from common causes of variation?

Failing to appreciate (and take coherent actions based on the understanding of) this difference, besides some lack of statistical knowledge, actually has to do with the very human, structural inability to deal with *uncertainty* and *decision making*, regardless of our cultural background or educational level. Behavioural psychology calls it "Cognitive Bias". In our jargon we refer to them as *Cognitive Constraints.*

In his popular show "Let's make a deal", Mr. Monty Hall provides a glaring example of how probabilistic uncertainty eludes the grasp of the human mind.

2.2 Uncertainty, Probability and the Monty Hall Problem

Monty Hall is famous for having hosted the television game show "Let's Make a Deal" and for the puzzle named after him: the very well-known *Monty Hall problem.*

The puzzle is posed as if it were a game on Monty Hall's show. Let's see how it works.

We have three doors, as in Fig. 2.1. Behind two of these three doors there are two things with little perceived value (for example, two goats) and behind the last door there is something of precious value (for example, a luxury car).

At this point, Mr. Hall asks the player to choose a door. Mr. Hall knows where the precious thing is. Therefore, regardless of the player's choice, Mr. Hall will open one of the two closed doors behind which there is a low-value thing (Fig. 2.2).

The player is now asked this question: do you want to change your choice?

When posed with this question the player's answer is *rarely* based on rationality. In the majority of cases, people would think that *there is no difference in probability* in finding the luxury car, whatever the choice may be.

Fig. 2.1 Three doors

Fig. 2.2 The player chooses a door

However, this is not true.

As you can see in Fig. 2.3, the probability of finding the luxury car doubles when you decide to change your original choice (we go from the initial 33–67%).

Being able to *calculate probabilities* is a big help as it allows us to deal with uncertainty and support a rational decision. Probably, the reader will still find this explanation puzzling because it is counter-intuitive. Indeed, thousands of people rejected the conclusion that changing the original choice was beneficial when the puzzle was published in a magazine (including many hundreds with a PhD). This inability to recognize a rational solution is an emotional blockage that we call *cognitive constraint*.

How can we deal with this?

We face similar cognitive problems when dealing with variation. The reason why we must acquire some basic statistical knowledge is to learn how to take *rational decisions* based on our ability to understand the nature of variation that affects our process/system. Eighty years of cumulative professional practice makes us very confident in saying that investing in the reading of the next few pages will not be a waste of the reader's time. It will equip them instead with the fortitude to embrace the new paradigm underpinned by Network Management.

In order to understand how the System behaves and how variation affects its functioning, we need a method to help us to discern the two kinds of variation the System is impacted by.

To our knowledge, nobody in the world has done more than Dr. Donald J. Wheeler to make the issue of variation clearly understood. What follows is our humble attempt to reproduce his teachings (and indulge in a little bit of math).

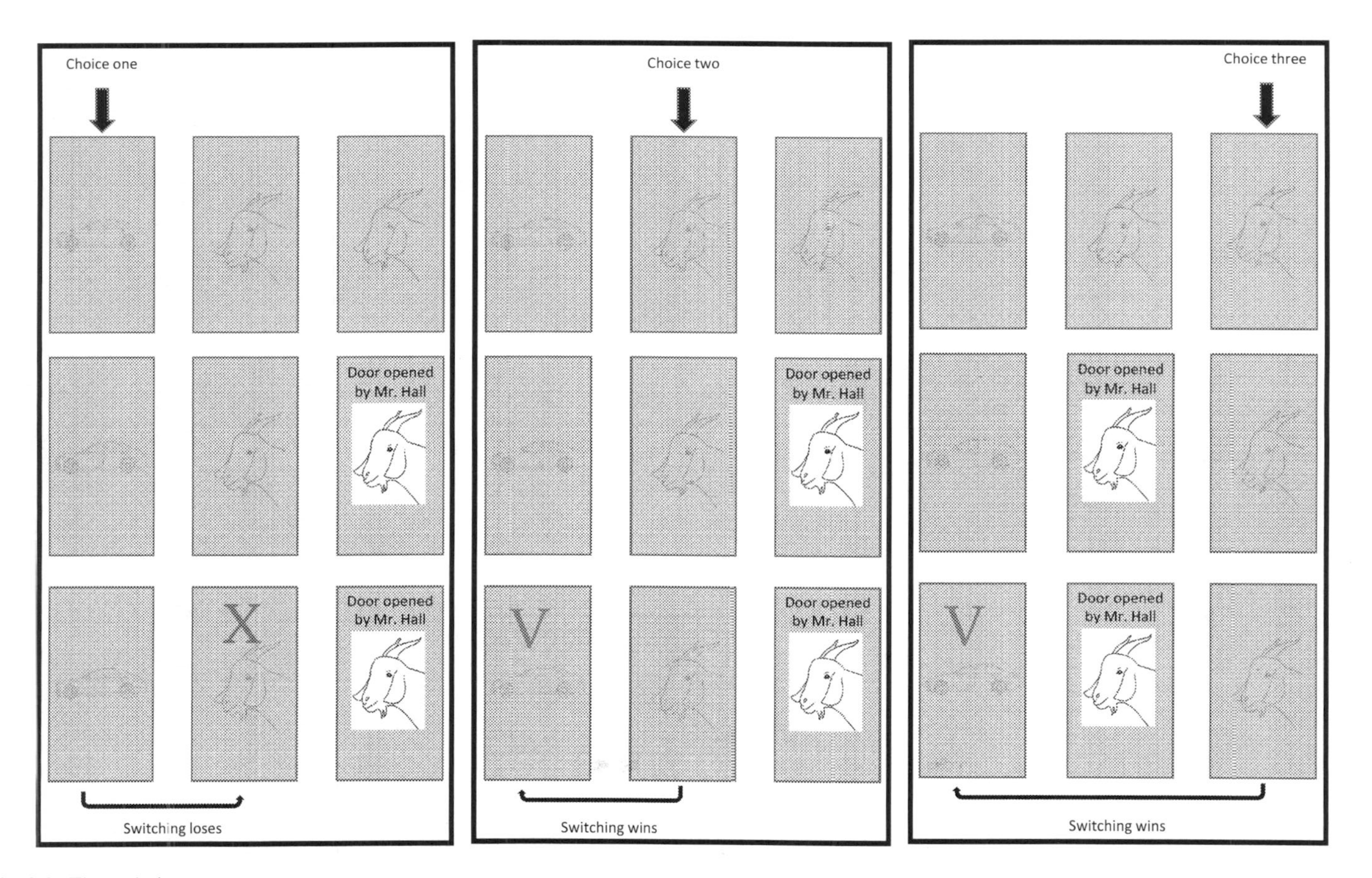

Fig. 2.3 Three choices

2.3 The Why and How of Process Behaviour Charts

Process Behaviour Charts (or Process Control Charts) are used to *unveil* the nature of a process, i.e. whether it is statistically *predictable* (we can predict its future behaviour from its past) or statistically *unpredictable* (we can say nothing about what is going to happen going forward).

We need the charts to avoid *unnecessary, undesirable* and often *fallacious* interference that may (and usually will) occur when managing a process, e.g. operators trying to keep the process "on target", rewarding or blaming individuals for sales performances, variance in Budget Versus actual analysis, etc.

Dr. Deming built upon this body of knowledge to formulate his systemic management philosophy and the *New Economics* that goes with it.

When we analyze a process and its variation, we deal with a set of data collected from a measurement. In order to understand the behaviour of the process, the first step we need to take is to understand the oscillation of the data around its average value.

The elements that form a Process Behaviour Chart are the set of data and three lines:

The Average of the process;
The Upper Natural Limit of Oscillation (Upper Control Limit);
The Lower Natural Limit of Oscillation (Lower Control Limit).

These Limits ARE NOT IMPOSED SPECIFICATION LIMITS. They represent the inherent, "natural" variation present in the data.

In formula form:

$$UNL = \overline{X} + 3\sigma_x$$

$$LNL = \overline{X} - 3\sigma_x$$

where σ_x is the standard deviation of the given data distribution. In essence, it is a measure of how the data are dispersed (distributed) around an average value.

The choice of $\pm$ 3σ for the upper and lower limits was chosen by Dr. Walter Shewhart based on *economical and empirical* considerations, *not* statistical ones. This is because the body of knowledge he developed while working at Bell Laboratories was aimed at solving an industrial problem; the monumentality of Shewhart's contribution lies in tackling a practical engineering issue with a scientific, epistemological, somewhat abstract, approach. In the 1920s, Shewhart laid the mathematical foundation for the management of Quality that Deming would go on to develop over the ensuing six decades.

What we show below is how to estimate the Standard Deviation.

Given a set of data, the AVERAGE identifies the 'centre' of the data:

$$\frac{\sum_{1}^{N} X_i}{N}$$

One common measure of how data are "spread" (distributed) around the average is the RANGE:

$$R = \max\{X_i\} - \min\{X_i\}$$

In order to give a reliable estimate of σ_x, we divide the data in subgroups and we use a more convenient measure of spread (dispersion); in the particular case of a *chart for single values* we use the MOVING RANGE (subgroup of dimension "2"):

$$mR = |X_{i+1} - X_i|$$

The standard deviation depends on the moving range as shown below:

$$\sigma_x = \frac{\overline{mR}}{d_2}$$

where d_2 is the statistical coefficient (bias correction factor) for the subgroup of dimension "2", whose numerical value is 1.128.

The limits for a chart for single values are then:

$$UNL = \overline{X} + 3\sigma_x = \overline{X} + \left(\frac{3}{d_2} * \overline{mR}\right) = \overline{X} + \left(2,66 * \overline{mR}\right)$$

$$LNL = \overline{X} - 3\sigma_x = \overline{X} - \left(\frac{3}{d_2} * \overline{mR}\right) = \overline{X} - (2,66 * \overline{mR})$$

UNL = upper natural limit.
$\overline{X}$ = average (mean).
LNL = lower natural limit.
$\overline{\text{mR}}$ = average moving range.

It is worth reemphasizing that the limits at ± 3σ *are not* probabilistic limits. There are some probabilistic considerations behind them, but Shewhart's decision for their choice was *empirical and economical.*

As a matter of fact, it had been known for over a hundred years, thanks to Gauss's work, that given a set of *homogeneous* data.

* About 60–75% of data will lie within ± 1σ from the mean.
* About 90–98% of data will lie within ± 2σ from the mean.
* About 99–100% of data will lie within ± 3σ from the mean.

Based on the above, Shewhart devised four *operational rules* to determine whether the behaviour of a process/system (its "condition") is predictable or not (whether it is, in a statistical sense, in a state of "control", or not).

1. There is evidence of an out-of-control condition every time a point falls beyond a limit (beyond $\pm 3\sigma$) as in Fig. 2.4.
2. There is evidence of an out-of-control condition every time two points out of three fall on the same side of the central line beyond $\pm 2\sigma$ as in Fig. 2.5.
3. There is evidence of an out-of-control condition every time four points out of five fall on the same side of the central line beyond $\pm 1\sigma$ as in Fig. 2.6.
4. There is evidence of an out-of-control condition every time at least eight consecutive points fall on the same side of the central line, as in Fig. 2.7.

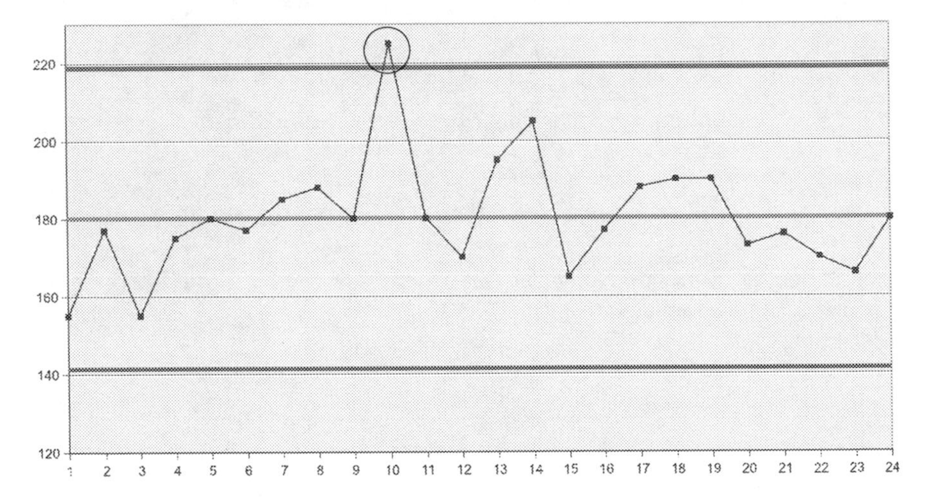

Fig. 2.4 Operational rule 1

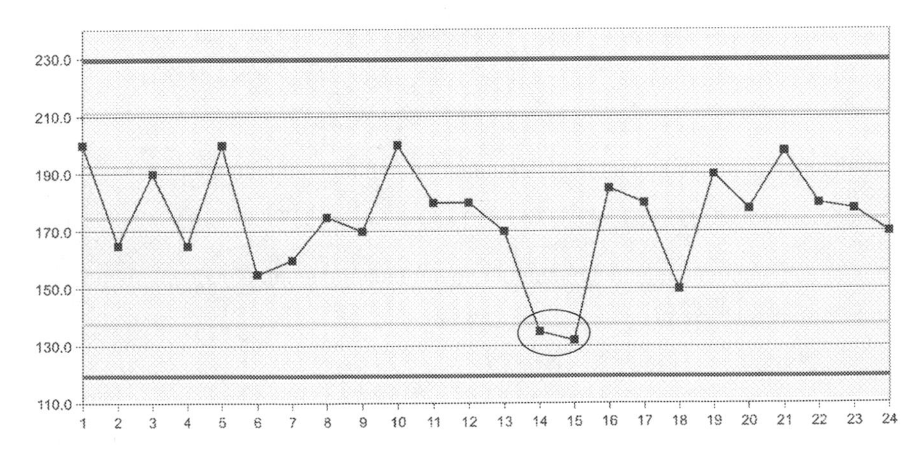

Fig. 2.5 Operational rule 2

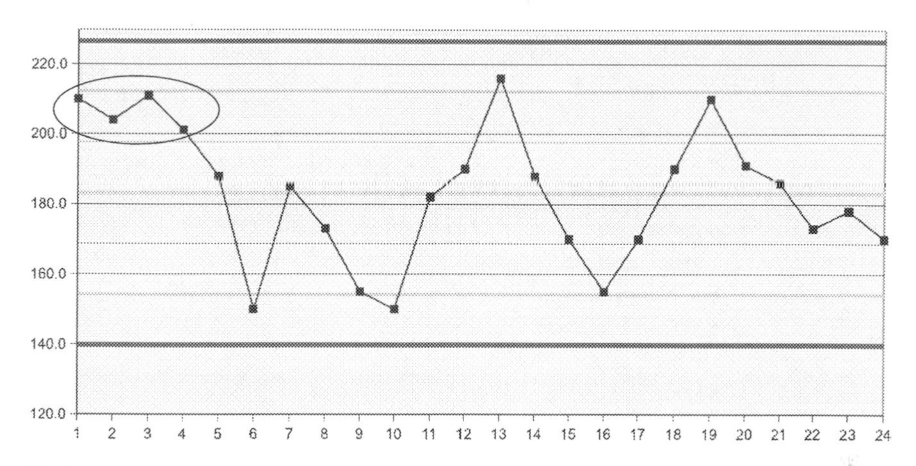

Fig. 2.6 Operational rule 3

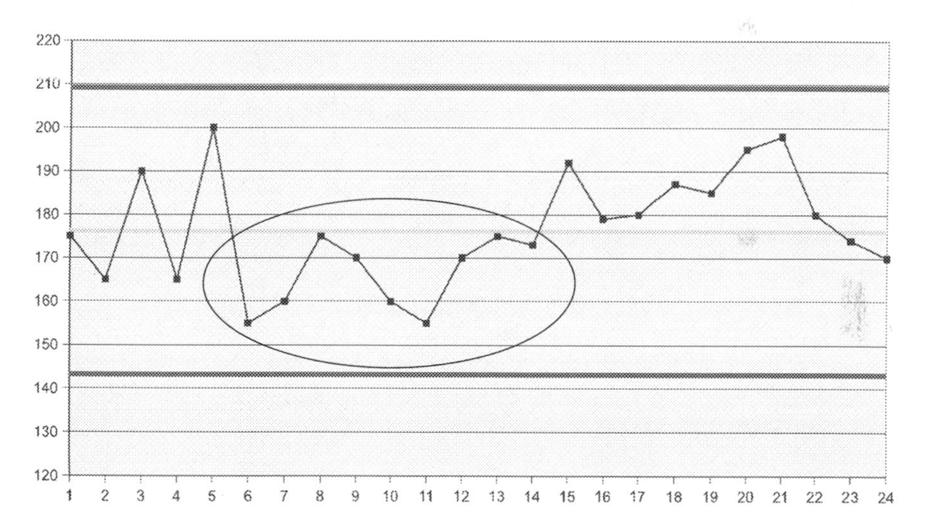

Fig. 2.7 Operational rule 4

The significance of the above rules is that the *probability* that any of those events occurs *by chance* is very small. There must have been a *special cause* for them to happen.

2.4 The Bias Correction Factor—Better Economics for the Standard Deviation

If the data we consider represents a set of *independent events* (meaning there is no correlation between two consecutive data measurements) the intrinsic variation associated to the process is just *random variation*. In this case, we can safely assume that the data distribute *normally* (Gaussian Distribution) and the range of variation (the dispersion around the average) is represented by the parameter σ (standard deviation) of the distribution. We approximate this by:

$$\sigma = \sqrt{\frac{\sum_1^n (x_i - \overline{x})^2}{n - 1}}$$

If the data represents, instead, a set of *dependent events*, two consecutive data measurements or groups of data can show a correlation/cause-effect relationship that can be disclosed by the appropriate reading of a Process Behaviour Chart; this is often what occurs in real processes.

In this case, the data might not distribute normally and we have to adopt a different approach that is valid for (almost) any kind of distribution.

Let's use a little theory as a guide for better practice. We assume that *routine variation* (intrinsic variation) is represented by a normal distribution (Gaussian).

We have a set of data $\{X\}$, for which the average (mean) and standard deviation are:

$$\mu = mean(X)$$

$$\sigma = SD(X)$$

Following Shewhart, we use a *local measure of variation* (to estimate the standard deviation in an easier way), *the range and its average.*

We arrange data in subgroups of size "n", according to the time sequence and chosen criteria.

We then calculate the average and the range for each subgroup in order to estimate the average and the standard deviation:

$$mean\left(\overline{X}\right) = mean(X)$$

$$\mathrm{SD}\left(\overline{X}\right) = \frac{SD(X)}{\sqrt{n}}$$

According to the *central limit theorem*, the distribution of the averages will approximate a normal distribution. In contrast, *this is not valid for ranges* and their distribution will be skewed.

$$mean(\mathrm{R}) = \mathrm{d_2(n)SD}(\overline{X})$$

where $d_2(n)$ is the *bias correction factor*, which depends on the size "n" of the sample.
To summarize:

$$mean(\mathrm{R}) = \mathrm{d_2(n)SD(X)}$$

The estimate of the standard deviation would be:

$$\mathrm{SD}(\overline{X}) = \frac{mean(\mathrm{R})}{d_2(n)}$$

Using standardized variables $\left(\text{standard deviation} = \frac{1}{\sqrt{2}}\right)$, we have:

$$mean(\mathrm{R}) = \frac{d_2(n)}{\sqrt{2}}$$

In conclusion, in order to determine the bias correction factor as a function of the size "*n*" of the sample, we have to calculate *mean* (R).

The range "R" is a variable for which we do not know the average, the standard deviation and the distribution. In cases like this, where all we have is an estimate from a small sample, the distribution used is the *t-Student* distribution, which is a function of the size of the sample.

This distribution approximates a Gaussian distribution with the increase of the sample size "*n*".

$$\frac{d_2(n)}{\sqrt{2}} = \overline{R} = \frac{1}{\sqrt{\pi}} \int_{-\infty}^{\infty} R_n F(R_n) dR$$

where $F(R(n))$ is a t-Student distribution:

$$F(R_n) = \frac{\Gamma\left(\frac{n+1}{2}\right)}{\Gamma\left(\frac{n}{2}\right)\sqrt{n\pi}}\left(1 + \frac{R^2}{n}\right)^{\frac{-(n+1)}{2}}$$

here "Γ" is the "gamma function" of Euler: $\Gamma(n) = \int_0^{\infty} e^{-x} x^{(n-1)} dx$ and:

$$\frac{d_2(n)}{\sqrt{2}} = \overline{R} = \frac{1}{\sqrt{\pi}} \int_{-\infty}^{\infty} R(n) \frac{\Gamma\left(\frac{n+1}{2}\right)}{\Gamma\left(\frac{n}{2}\right)\sqrt{n\pi}}\left(1 + \frac{R^2}{n}\right)^{\frac{-(n+1)}{2}} dR$$

the solution of this integral is usually numerical, but for n = 2 we obtain:

$$d_2 = 1.128.$$

2.5 Some Final Considerations

At this point, we know how to calculate the limits of the process and we have built trust in where the coefficient to estimate the standard deviation comes from.

The question now becomes: how many data points should we use to calculate the average and the average moving range? Depending on the process, a number between 15 and 20 data points is generally accepted as a good choice. (You can trust us on this or refer to the distinguished treatises by Dr. Wheeler listed in the bibliography).

But why can't we use more points? (For example, if the process is the injection moulding of a small cap where many items are produced per hour and we want to monitor the production yield, why can't we use 100 points or more?).

Let's remember that variation exists. If we leave a System to itself long enough the effect of *entropy* will become tangible and variation will definitely increase. Given enough time, *anything that can happen will happen* (this is what some people, irritatingly, call *Murphy*). If we use too many data points to calculate the limits, everything will become *routine variation* and we might mistake "faith" for "coincidence", or "noise" for a "signal".

A Shewhart (Behaviour) Chart compares *local variation* at each point in time with *global variation*; in this way we can make the assumption that local subgroups contain routine variation, but are largely uncontaminated by special causes of variation over time. This approach allows us to sift "signal" from "noise", and helps us to detect special causes of variation over time.

There is no "short term" or "long term". We simply have to consider the time interval over which the data is collected. There are no probabilities, no hypothesis tests, no Z shifts, no bell-shaped distributions, and no need to resort to The Central Limit Theorem to explain and justify the use of Behaviour charts.

Unfortunately, people fail to understand that Behaviour Charts *are not* probability charts and they do not depend on any probability model (despite being based on probability). With the increase in calculation capacity, it is possible to demonstrate that we do not need to make a hypothesis about the distribution of the original data. The above considerations are valid for almost 1200 different kinds of distributions.

2.6 Examples of Process Behaviour Charts

Let's give another blow to our conventional, linear thinking; this is important so we can use Buffers in a more proficient way.

Two examples.

(1) A production process.

The following two charts represent the same process with different output.

In Fig. 2.8, the process has an average of 150 and oscillates between 135 and 165. In Fig. 2.9, the process has an average of 178 and oscillates between 143 and 210.

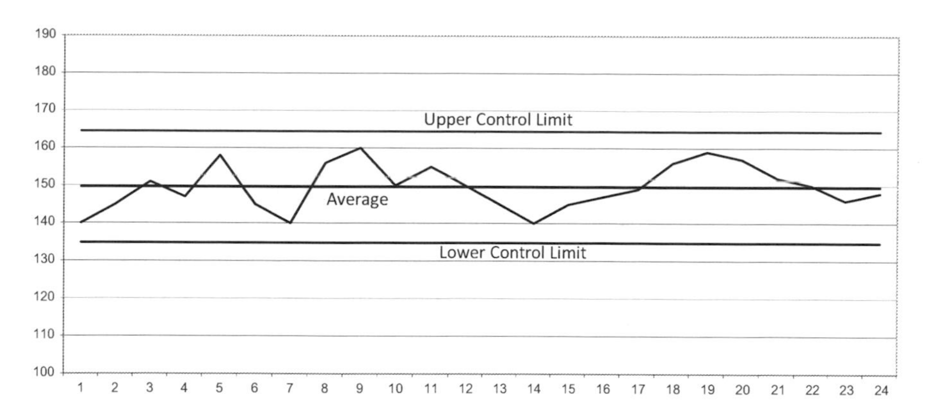

Fig. 2.8 A predictable process

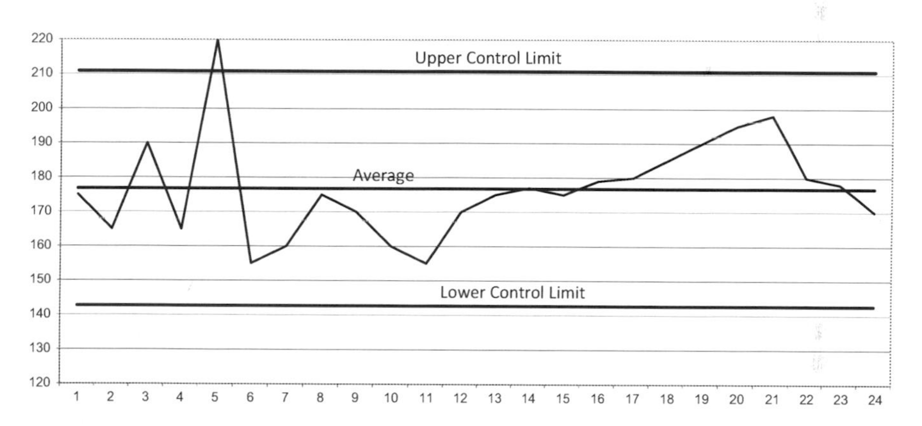

Fig. 2.9 A process with no inherent predictability

While the average value of the latter might be, in principle, more desirable (higher average output value), the range of oscillation (variation) is much wider than the former (30 against 67) *and* we can immediately identify one point in the second chart that is beyond the upper limit of variation, there is no inherent predictability in this process—anything could happen tomorrow.

Let's keep in mind that the UCL (Upper Control Limit) and the LCL (Lower Control Limit) are *calculated*, just as the average is. They *are not specifications* and do not represent the maximum and the minimum of the data series. They represent *the voice of the process,* i.e. *the natural (intrinsic) variation of the process.*

What we can say about these two processes is that the former exhibits some level of consistency over time and we can say something about its future; about the latter there is nothing we can say, whether its outcome is desirable or not.

(2) Arrival time at work.

The following two charts represent the arrival time at work for two different workers (who are supposed to clock in at 8 am).

The first worker in Fig. 2.10 shows an average arrival time of 8:01, and the oscillation is between 6:15 and 9:48 (pretty bad, one may say).

The second worker in Fig. 2.11 shows an average arrival time of 7:57, and the oscillation is between 7:38 and 8:15 (pretty good, one may say).

Again, we can immediately identify one point beyond the upper limit of variation in the process for the second worker. We cannot say anything about what his/her next arrival time will be.

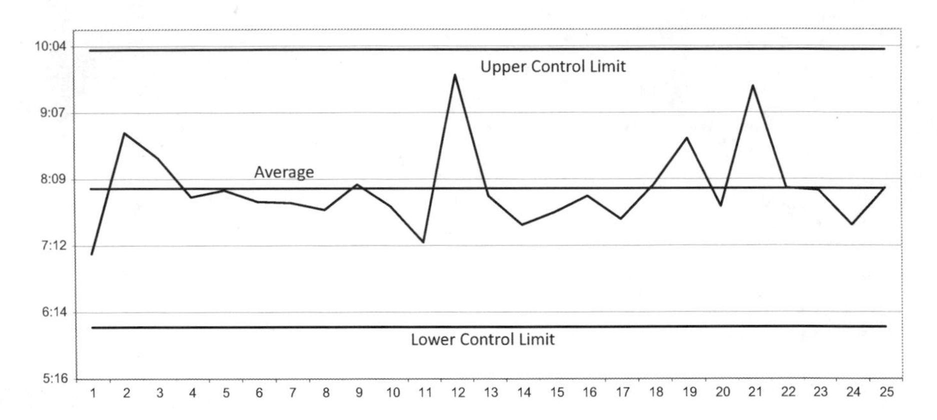

Fig. 2.10 A high variation process

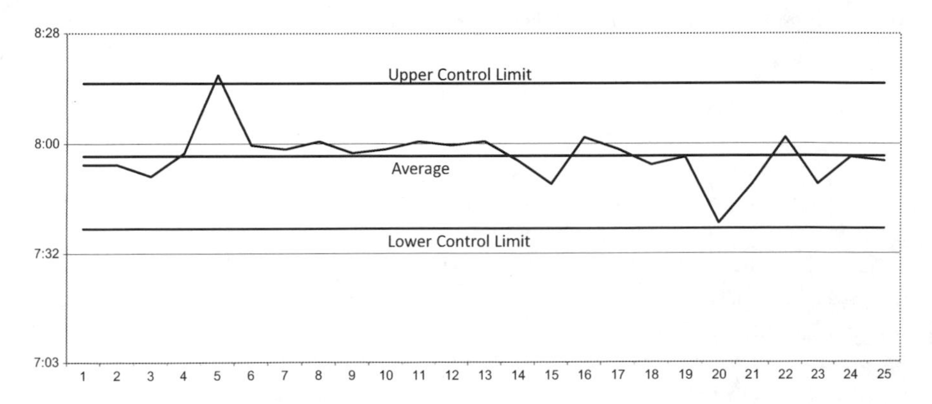

Fig. 2.11 A low variation process

In the first example we have.

Case 1—*low variation* and process oscillation within its natural limits *(in statistical control).*
Case 2—*high variation* and process *out of statistical control* (one point beyond the UCL).

In the second example we have.

Case 1—*high variation* and process oscillation within the limits *(in statistical control).*
Case 2—*low variation* and process *out of statistical control* (one point beyond the UCL).

The lesson is: a wide range of variation does not imply that a process is necessarily unpredictable ("out of statistical control"). Similarly, a narrow range of variation does not imply that a process is predictable ("in statistical control"). Very different managerial decisions must be taken to address these four situations and we highly recommend undertaking further studies as the success of a synchronized Network of Projects will rely heavily on these decisions.

All of the above is, evidently, counterintuitive and it does require some sort of "re-wiring" of our thought process. This is why we need these charts. Simply speaking, the human mind struggles with statistical thinking.

In short: relying solely on intuition (e.g., high variation means unpredictability and vice versa) to make decisions is misleading. Actions following *only* intuition in a complex environment lead us to wrong decision-making triggered by both ignorance and the ever present *cognitive constraints (cognitive biases).*

2.7 Variance Versus Covariance: The Role of Buffer Management

Let's now learn how to use a Buffer as an effective way to manage the complexity of a Network of Projects and control its evolution over time.

We have defined the Constraint of a System as *the element that determines/dictates the pace at which the System generates units of the goal.*

In a system that is "unbalanced" around, and subordinated to, the constraint, it is imperative to protect the constraint from variation that may disrupt its functioning. We call this protection *Buffer.* The Buffer protects the constraint from the cumulated variation (roughly speaking, *Covariance*) in the whole system. In a somewhat imprecise but functionally correct way, we can say that the Buffer protects against the "Co-Variation" in the system precisely where it would hurt the most, on the constraint (see Fig. 2.12).

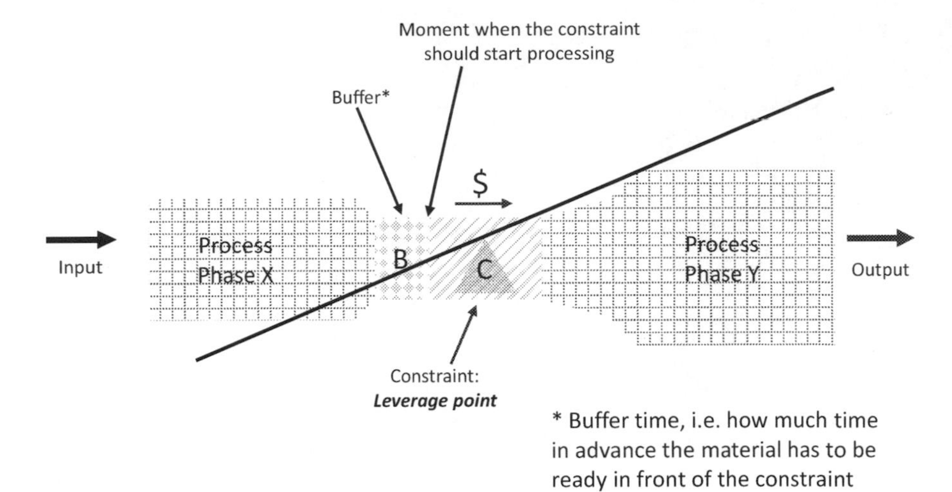

Fig. 2.12 The buffer protects against the effects of "co-variation" on the constraint

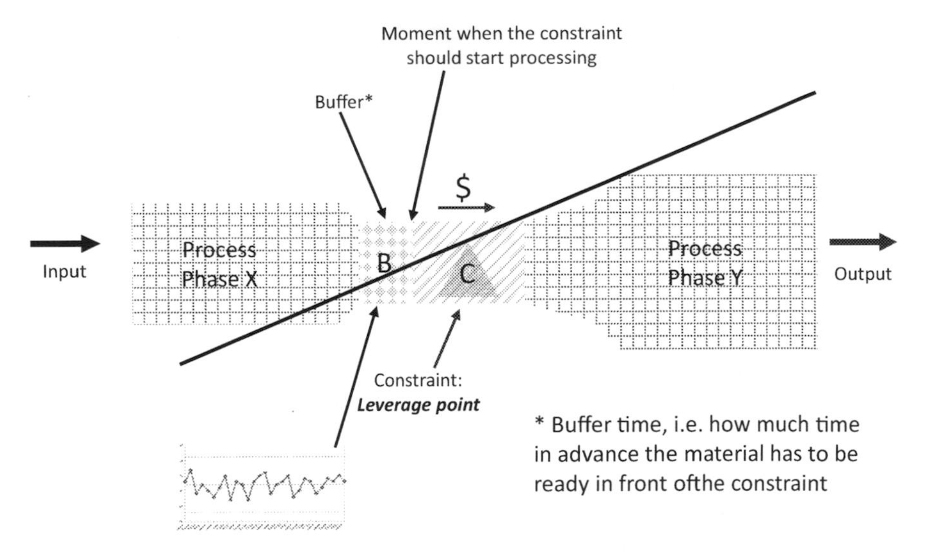

Fig. 2.13 Monitoring the variation of the co-variation

By monitoring variation in buffer consumption (with a Process Behaviour Chart as in Fig. 2.13) we get a good handle on what is going on throughout the whole system. Essentially, by monitoring "the Variation of the Co-Variation" we have a firm grip on how the whole system is performing, i.e. *predictably or not.*

Indeed, monitoring is not the exercise of an *inertial observer* and managers are not *spectators* in the organization.

The *dilemma* managers constantly experience in the face of processes perceived as "out of wack" (agreed, not a very scientific terminology…) is "intervene versus not intervene". The assumption underlying this dilemma is that *"there is no cut and dry rule for intervention (Gut rules)"*.

As we will see, the solution (injection) to this dilemma is to *define clear rules of intervention on the basis of buffer consumption* (see Appendix C).

It is interesting here to see how the *empirical* rule of intervention, initially developed by Dr. Goldratt, gives only a partial answer to the problem; it is one that inherits the hard to dispel confusion between natural oscillation limits and imposed specification limits. Time for a closer look.

2.8 The Three-Zone Approach

In a typical situation of a production flow, an ideal scenario is that the complete set of items the constraint has to work on (batch) show up in front of the constraint "one buffer time ahead". The traditional Theory of Constraints approach divides the Buffer into three zones, as in Fig. 2.14, and suggests different managerial behaviours according to the zone of the buffer that has been 'penetrated'.

The buffer will "empty out" or "fill up" depending on whether the tasks/processes leading to the constraint are completed early or late in respect with the scheduling, creating so called "holes in the buffer".

Unfortunately, this approach does not consider the most important aspect of variation: its *nature.*

The epistemological stance that systems management is rooted in dictates that *the essence of management is prediction*; only if we are able to foresee the outcome of our actions on the system can we make decisions that anticipate the development of events and therefore have more control over those events.

The real question that should dictate management behaviours becomes: "Is the system statistically stable?" (i.e. predictable).

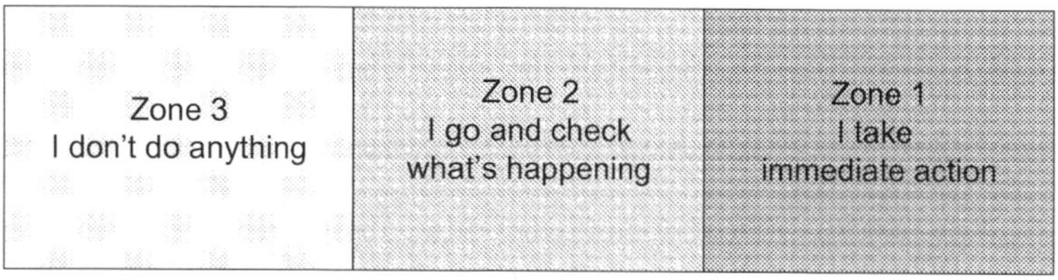

Zone 3 I don't do anything	Zone 2 I go and check what's happening	Zone 1 I take immediate action

Fig. 2.14 Three-zone approach

2.9 The Price We Pay for Ignorance

Let's take a look at four different scenarios.

In the first case, we detect 'holes' in the buffer in zones 2 and 3, as in Fig. 2.15. The three-zone approach leads us to act on the process (go and check). As the consumption process is statistically stable and oscillates within the three zones, we don't need to take any action (not even "go and check").

In the second case, we detect holes in the buffer in zones 2 and 3, as in Fig. 2.16. The three-zone approach leads us to act on the process (go and check). However, the process is now out of statistical control, i.e. not predictable. Even if all the points fall within zones 2 and 3, we need to act *immediately* and get the process back in statistical control (*work on the System*).

In the third case, we detect 'holes' only in zone 3 of the buffer, as in Fig. 2.17. The three-zone approach prescribes not to intervene on the process (and may even suggest to shorten the buffer); however, the process is *out of statistical control* (i.e. completely unpredictable). Therefore, we need to act immediately just as in case 2; we must bring the process back into statistical control (*work on the System*).

In the fourth case, we detect 'holes' in the buffer in zones 1 and 2, as in Fig. 2.18. The three-zone approach urges immediate action and leads us to *interfere* with the process (tamper with the System). Yet, given the stability of the process there would be no reason to interfere. As this process is predictable with an upper limit completely within zone 1, we do not need to take any action.

One final remark on the sizing of the buffer: if the individual and cumulated variation of the processes that impact the buffer are not statistically stable, we cannot knowledgeably define the amount of protection needed (size of the buffer).

The buffer acts as an effective means of control *only* if the variation in the processes of the system (and their cumulated effect) are in a state of *statistical control*. If

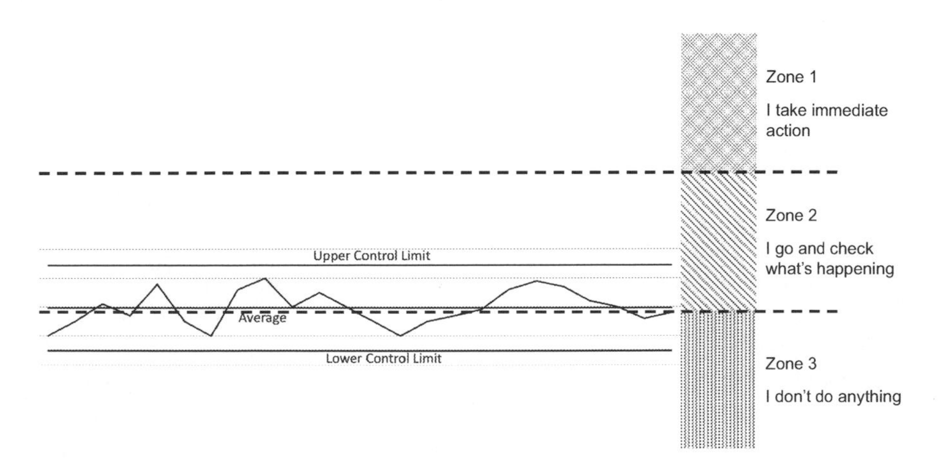

Fig. 2.15 Holes in the buffer in zones 2 and 3

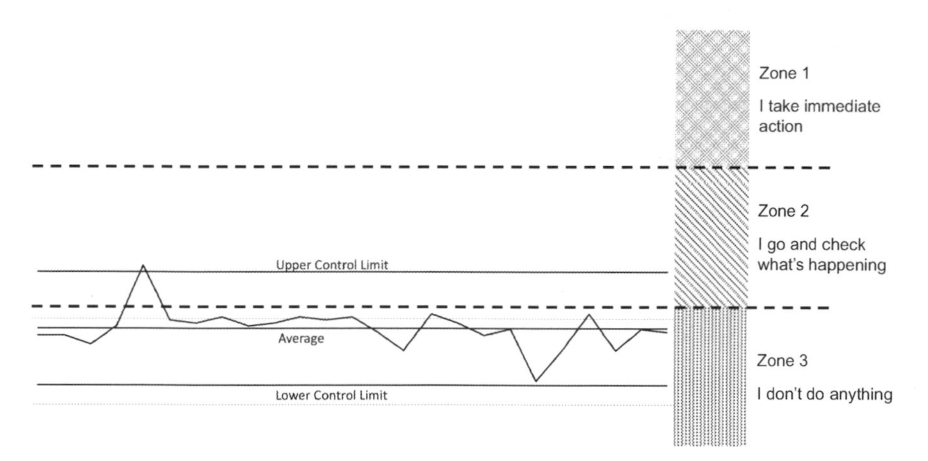

Fig. 2.16 The process is out of statistical control

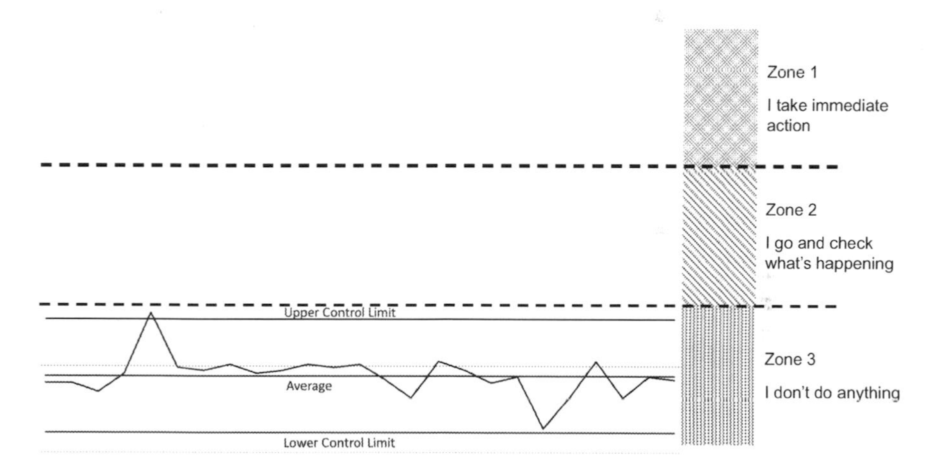

Fig. 2.17 Holes in the buffer only in zone 3

this is the case, to ensure sufficient constraint protection the oscillation of buffer consumption must show an upper limit that is lower than the size of the buffer.

Buffer Management means, in essence, understanding how variation affects what the organization does to achieve its goal. Understanding variation then becomes key to a successful transformation from Silo to Network through synchronizing a Network of Projects. Chapter Three of this book is dedicated to describing operationally the algorithm that enables this transformation.

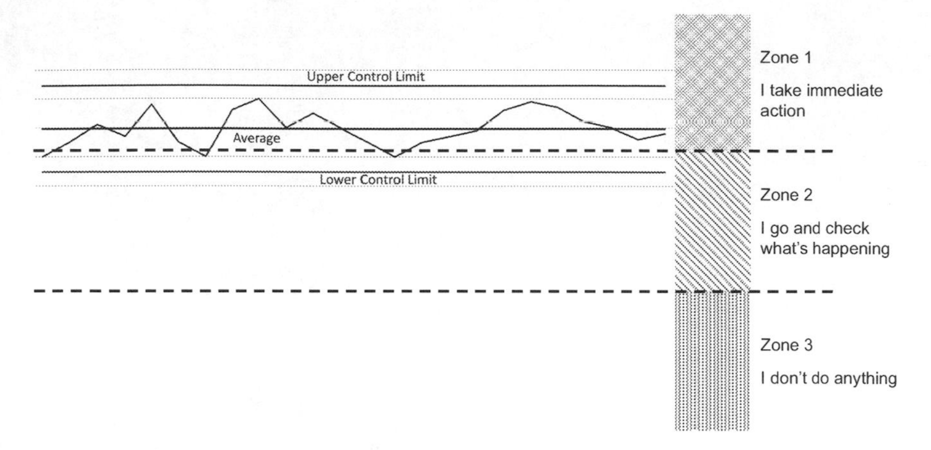

Fig. 2.18 Holes in the buffer in zones 1 and 2

2.10 A Very Brief Outline on Network Theory

A network is a set of elements consisting of *nodes* and their connections, referred to as *links*. The links are not only a physical connection, they represent the *relation* between nodes.

A network approach focuses all the attention on the *global structure* of the interactions within a system. As a matter of fact, we disregard the individual properties of each element by itself. As a result, systems are all described by the same tool: a *graph*.

A "random graph", or "random network" can be simply defined as a network where the distribution of links between nodes follows a *Poisson Distribution,* or a *Gaussian Distribution.* In this case, each node of the network has a high probability of having a number of links that is close to the average of the Distribution.

A "scale-free network" is, instead, defined as a network where the distribution of links between the nodes follows a *Power Law Distribution.* In this case, there will be a concentration of links at certain nodes, which are called *hubs***,** while the number of links decreases according to a *power law* for the rest of the nodes.

Two typical examples of these kinds of networks are shown below.

(1) The picture in Fig. 2.19 represents the road network in the USA.

This is a very well-known example of a random network. As you can see, on average we have the same number of links at each node (a number which is equivalent to the average of the distribution).

(2) Conversely, the picture in Fig. 2.20 represents the flight paths in the USA.

In this case, the number of links is higher at certain nodes (the air hubs), generating a distribution of links that follows a power law.

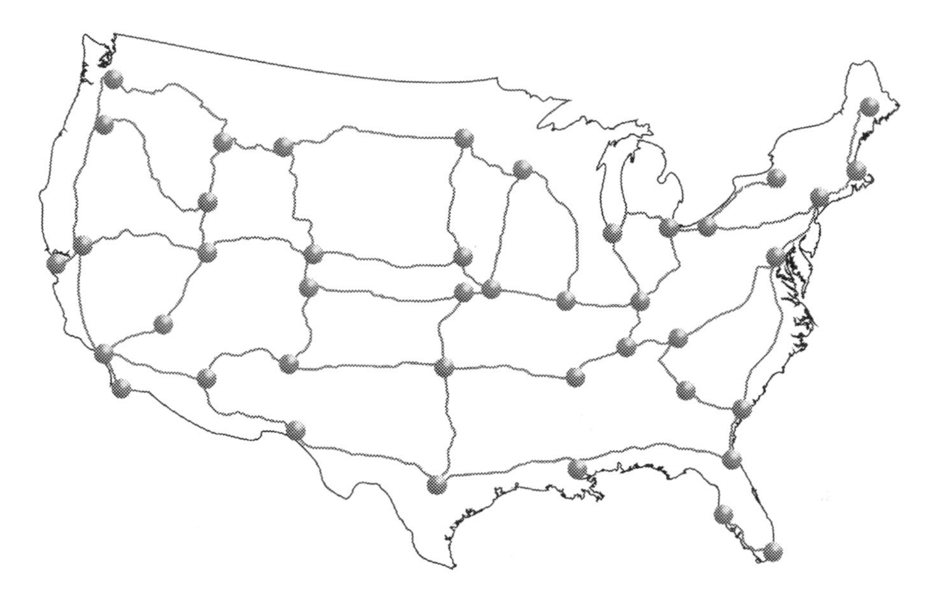

Fig. 2.19 Representation of the road network in the USA

Fig. 2.20 Representation of flight paths in the USA

In practical terms, the main difference between the two kinds of networks lies in the way they grow and the way they function. Due to the random distribution of links among nodes, the possibility that an accident could create a disruption in the USA road network is remote. On the contrary, due to the concentration of links at certain nodes, it is likely that an accident to any of the hubs in the flight paths network would create a disruption to the entire system. In conclusion, a random network is "more solid" or resilient compared to a scale-free network.

2.11 Multi-project Environments and Buffer Management in the Network of Projects

We should highlight some of the implications of managing the System in a multi-project environment, or what we also refer to as a *Network of Projects.*

The Network of Projects is an *Oriented Network* (4), meaning that all the efforts from the different elements of the System (competencies, resources, machines, etc.) must be oriented toward the goal of the System. According to this definition, all the projects in question are *interdependent.*

Let's try to understand better what kind of network we are dealing with.

Each task in the network (node of the network) is characterized by:

- A time duration;
- A resource, or rather, a level of competency associated with a resource.

The Critical Chain is a *logical network* which defines the length of the project. The links between tasks—also called nodes—of the Critical Chain are logical links, not physical links.

In order to obtain the Critical Chain, the algorithm *maximizes* the length of the chain of interdependent tasks which exists both among tasks in a single project and among tasks from different projects, after solving the resource contention. This process determines the logical attachment among tasks.

In any project network, the tasks which will show most connections are those belonging to the Critical Chain. In particular, these tasks will be those with the *most critical* associated competency.

The growth of a "Network of Oriented Projects" follows a *scale-free* network model, where the *hubs* are the different Critical Chains, and within each Critical Chain the hubs are the tasks on the Critical Chain (resembling a kind of "self-similar fractal" structure). This is why, in order to have a more "solid" network of projects, we have to make sure that we have more resources with different competencies. Ideally, we should make sure that the network of projects is a *as random a network as possible.*

What we mean by this is that if we had "infinite competencies" associated to infinite resources, we could generate a perfectly random network of projects with no tangible limitation. In a network with infinite competencies associated to infinite resources we can manage as many projects as we want.

We feel that one final remark is necessary. There is an important difference between a network as it is described in any book on Network Theory and the Network of Projects we have just delineated. While networks evolve according to the dynamics of a self-organized system, the Network of Projects is governed by the dynamics of a system which is *oriented toward its goal* (a goal which is, in turn, determined by the constraint).

2.12 The Effect of Complexity on the Network of Projects

In managing such a Network, we always have to take into consideration that *any* delay in one of the projects of the network will automatically be transferred to the other projects of the network, as in Fig. 2.21.

In this example, the late completion of Task 1 in Project 2, where resource R4 is involved, will impact Task 2 on Project 2 where resource R3 is involved.

As a direct consequence of the *interdependence of the projects*, Task 3 in Project 1 cannot start because resource R3 will still be busy completing Task 2 in Project 2. *Complexity* emerges naturally if we extend the above reasoning to all the tasks of the different projects of the network. The manifestation of complexity becomes evident in what we call the *butterfly effect*.

The only way to minimize the impact of the butterfly effect is to:

- assign the right level of competency to each task;
- standardize, as much as we can, the task execution in order to increase the statistical predictability of the completion time.

The impact of delays, which will be unavoidable because of *routine variation*, will be absorbed by the appropriate sizing of the buffer.

Regarding the activity of *Buffer Management* in Project Management, one more clarification is necessary. Generally speaking, routine variation can have both a *negative* impact or a *positive* impact on the buffer size. The buffer size oscillates as a consequence of the positive or negative impact of routine variation. This is valid for the buffer in front of a *physical constraint*. In Project Management, the buffer is positioned at the end of the Critical Chain and, because of the intrinsic nature of project execution, the consumption of the buffer can only increase. The benefit of positive variation (finishing a task early) will be lost, while negative variation (finishing a task late) will be transferred to the chain.

This fact makes the use of the traditional Behavior Chart for monitoring the buffer oscillation less effective. For this reason, in order to perform the activity of Buffer Management, we use a different kind of Behavior Chart: a *Trend Behavior Chart*.

A Trend Behavior Chart is made up of the three straight lines below, plus the series of data representing the buffer consumption:

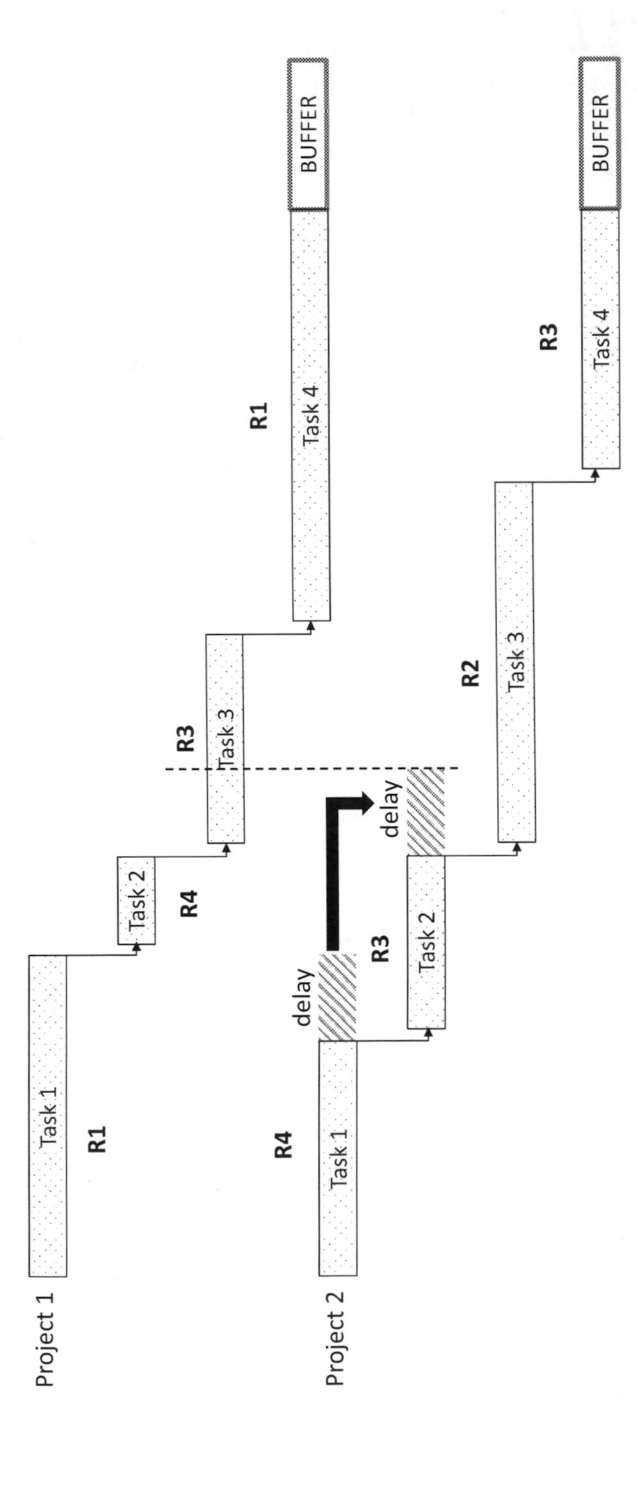

Fig. 2.21 Task 1 is completed late

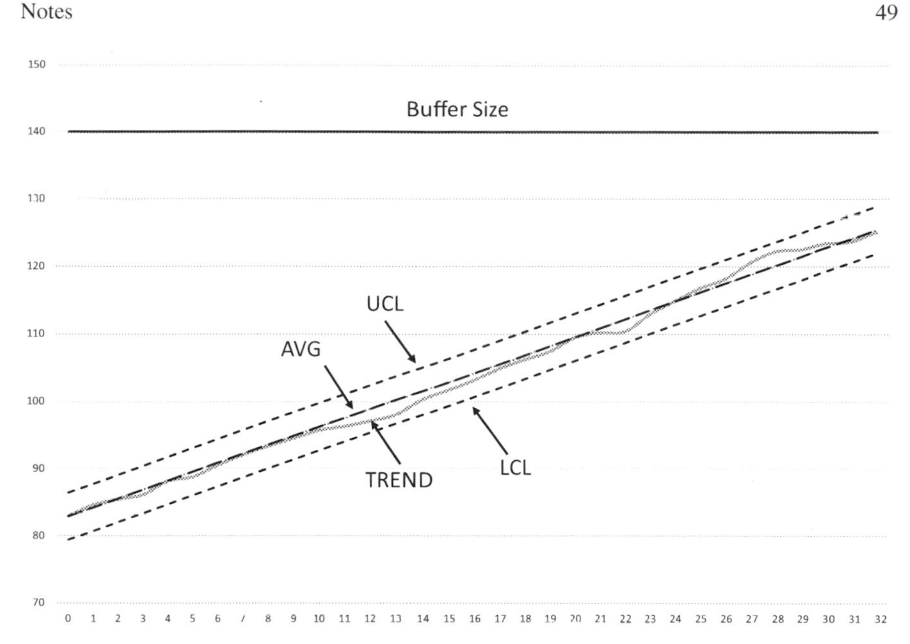

Fig. 2.22 A trend behavior chart

- the data series;
- the average trend;
- the Upper Control Limit (a straight line parallel to the average trend at a distance of $+3\sigma$);
- the Lower Control Limit (a straight line parallel to the average trend at a distance of -3σ).

This chart will give us information about how predictable the rate at which we use the buffer is.

Since the buffer can only be consumed, if the rate of consumption is statistically predictable (within the Upper and Lower limits), we will know if, and when, the buffer will potentially be emptied, as in Fig. 2.22.

Notes

1. Goldratt (1997).
2. See: Lepore et al. (2016).
3. Shewhart (1931).
4. We discuss the nature of networks in the context of their relationship to organizations, variation and our organizational design *The Network of Projects* in our chapter 'Making the Change Operational: The Network of Projects' in Lepore (2017).

References

E.M. Goldratt, *Critical Chain* (North River Press, Great Barrington, MA, 1997)

D. Lepore, A. Montgomery, G. Siepe, Managing Complexity in Organizations Through a Systemic Network of Projects, in *Applications of Systems Thinking and Soft Operations Research in Managing Complexity*. ed. by A. Masys (Springer International Publishing, Switzerland, 2016), pp. 35–69

D. Lepore, A. Montgomery, G. Siepe, *Quality, Involvement, Flow: The Systemic Organization* (CRC Press, New York, 2017)

W.A. Shewhart, *Economic Control of Quality of Manufactured Product* (van Nostrand Company Inc., New York, 1931)

Chapter 3
The Ess3ntial Algorithm

Abstract We introduce and describe in detail the structure of the algorithm for Ess3ntial–a systemic solution for managing and synchronizing multi-projects at finite capacity. We present a new way of considering available resources as a pool of competencies and how these competencies can be scheduled instead of resources, introducing a far greater flexibility in the ability of an organization to manage several projects in parallel and monitor and manage their progress.

Keywords Critical chain · Multi-project environment · Competencies · Project scheduling · Buffer management

3.1 Introduction

As we have stated, an organization is a system, a network of interdependent components that work together to achieve a common goal. We can manage an organization as a network of projects, where we understand that each project is a *network of interdependent activities* that must be completed within a given timeframe with a certain number of resources.

With *scheduling* a project we mean the procedure whereby we organize and assign the resources available in the system to the various phases. This is generally rather complex, especially as the number of resources available is naturally finite and a project may require various tasks to be carried out simultaneously. For this reason, a method must be adopted to optimize the way that resources are assigned to tasks and this can be done on the basis of a wide variety of criteria.

A range of methods exist for scheduling projects with varying levels of effectiveness and some software products are available to support implementation. The algorithmic methods underpinning various kinds of scheduling can be very different, depending on the approach that is adopted. Among the software solutions that have been made available, *Independence* was the first software to accompany The Decalogue management methodology. It was created in the late 1990s.

In this book we focus on the algorithm that is the engine of the Ess3ntial software. Ess3ntial is both the offspring and a further evolution of Independence as it takes

D. Lepore et al., *From Silos to Network: A New Kind of Science for Management*, SpringerBriefs in Complexity, https://doi.org/10.1007/978-3-031-40228-9_3

into account the development of the Decalogue methodology thanks to the interim years of implementations internationally in increasingly complex environments since the mid 1990s. Like Independence, Ess3ntial follows the general guidelines of the *Critical Chain* method for project management developed by Dr. Eliyahu Goldratt.

Compared to other software products available for managing projects with Critical Chain, Ess3ntial introduces substantial innovations and it revolutionizes the approach to scheduling projects in several ways.

3.2 Competency Based Scheduling

The main innovation that Ess3ntial brings is to measure the system's capacity in terms of *competencies* rather than resources available. This allows an organization to manage a pool of competencies instead of a set of resources. This shifts the focus away from individual resources towards the competencies those resources possess. This is a true innovation in project management as it introduces a new level of flexibility in the scheduling phase based on the number of competencies available overall rather than the number of people.

Moreover, *assigning resources* to tasks can be made completely automatic, according to well defined optimization methods. Once the competencies that are necessary to carry out a task have been identified, the software selects the resources who are in fact available based on their calendars with the appropriate competencies, starting with those who display lower criticality in terms of a series of pre-defined indicators within the algorithm.

Currently, resources are mostly assigned manually by project managers and these choices are often emotional rather than driven by scientific/rational criteria. This kind of assignation is based on the assumption that a given task must be carried out by a particular person rather than by a specific competence that more than one person might possess.

Insisting on a specific resource to carry out one or more tasks might in fact create delays and conflicts that could slow down the completion of projects. While Ess3ntial has been conceived to take advantage of the availability of competencies, it is also possible to assign manually a task to a specific person.

3.3 Measuring Progress

We can always know how a project is progressing and if any delays are accumulating by checking on *how much the project buffer is being consumed*. This measurement is supported through the use of Statistical Process Control, first introduced by Walter Shewhart. This is a more scientific and insightful way of measuring buffer consumption than the conventional "three zones" in the Theory of Constraints.

Moreover, the control chart of buffer consumption enables a timely planning of corrective action.

3.4 Multi-project Environment

Ess3ntial is a truly *multi-project solution* (as opposed to the multi single-project solutions found in other Critical Chain software) as it allows project managers to assign resources without creating any resource contention even when several projects are taking place simultaneously. This is of fundamental importance when we want to manage an organization systemically as a *network of projects.*

The main difficulty in managing several projects simultaneously lies in resolving the contention of resources for the various tasks in different projects. This is an aspect that is generally highly underestimated in other algorithms on the market for scheduling and managing projects.

3.5 How the Algorithm is Structured

In this chapter we provide a detailed description of the algorithm that is structured around the following phases:

- Definition of the variables that are indispensable for implementing the scheduling algorithm; for this we assume that we have analyzed the system in depth and classified the data necessary for our objective. In particular, we must know the *resources of the system,* and the *competencies they possess* (*variables of the system*). To schedule a project we must identify the *variables of the project*. In particular, we need to know the *tasks* of the project, the *estimate of duration for each task* and the *list of competencies necessary* to carry it out. Moreover, we must establish the connections among the tasks that constitute the *project network.*
- We then move on to the actual scheduling phase. This entails identifying the steps necessary to create the network of the project and building *the critical path,* i.e. the longest logical sequence of tasks that represents the backbone of the project.
- Next, we tackle the rational assignation of available resources to the tasks of the project. In particular, we will see which criteria are used to choose the most suitable resources from among those available within the system to carry out a given task.

  Moreover, we will illustrate how to tackle (and possibly resolve) any resource contention among tasks to be carried out simultaneously. Starting from the critical path, this phase ends with the identification of *the critical chain,* i.e. the longest sequence of interdependent tasks, taking into account both logical dependencies and the sharing of resources.

The critical chain is the physical constraint of the project as it determines the real duration of the project.

Finally, scheduling is completed with the insertion of the *project buffer*: the project buffer protects the critical chain from any delays in execution and *feeding buffers* protect secondary branches in the project that feed into the tasks along the critical chain. This is necessary because any delay in executing these tasks could impact the critical chain, thus causing it to lengthen.

- The following phase is the description of the methods used to monitor the state of progress of an ongoing project by measuring consumption of the project buffer. In particular, we will see how the Ess3ntial algorithm uses a dynamic method to measure project buffer consumption that is initially based on the visualization of three zones in red, yellow and green that, like a traffic light, indicate progressive levels of buffer consumption (this same method is used to measure the consumption of feeding buffers). 15 days after the beginning of a project, there will be sufficient data to continue monitoring the buffer with a *control chart*.
- Finally, we will describe the multi-project function, i.e. the way that the Ess3ntial algorithm manages the execution of two or more projects simultaneously.

3.6 Variables

We will introduce the variables that are indispensable for scheduling and we will separate them into two categories: variables of the system relevant for any project and specific variables of a project.

We have used the language of matrices for the mathematical definition of the variables.

We will assume that the following general information about our system is known:

- The list of the N resources available in the system, listed in the form of N-dimensional vector:

$$R = \{r_i\}_{i=1,\dots,N} = (r_1, r_2, \dots, r_N)$$

- The list of the M competencies available overall:

$$C = \{c_i\}_{i=1,\dots,M} = (c_1, c_2, \dots, c_M)$$

- The list of the competencies possessed by each resource r_i, gathered into a Boolean matrix:

$$K = \{k_{ij}\}_{i=1,\dots,N\ j=1,\dots,M}$$

where we assume that the element k_{ij} of matrix K, with N rows and M columns, can take on the following values:

$k_{ij} = 0$, if resource r_i does **not** possess competence c_j.
$k_{ij} = 1$, if resource r_i possesses competence c_j.

The value k_{ij} is in fact *retrieved* from one of the values from the M-vector C, in such a way that, for example, if $k_{32} = 1$, it means that the resource that appears at number 3 in the list of resources of the system (i.e. resource r_3), possesses competence number 2 in the list of competencies (i.e. competence c_2).

- The list of calendars showing when resources are busy:

$$W = \{w_i(j)\}_{i=1,\dots,N, j\geq 0}$$

At the start, if no project has yet been scheduled, we assume that $W = \emptyset$, i.e. that no resource is busy. Each element $w_i(j)$ consists of an interval of time that is measured in days (or hours of work).

The index j indicates the number of tasks the resource r_i is allocated to; it is increased each time the resource is assigned to a new task.

In order to create a project we must know all the information that makes scheduling possible. We will use the following:

- The date E when we intend to complete the project.
- The amount of protection we wish to give the project in terms of a percentage of the duration of the project (project buffer): *pb*.

  In general, this is a choice the project manager makes and they may decide to assign a certain percentage of buffer to each project. In some cases, depending on the nature of the project in question, the project manager can decide to add more or less protection through the size of the buffer.

  This entails introducing a project buffer as an extra task at the end of the critical chain, the duration of which is a percentage *pb* of the critical chain.

  Moreover, as all the non-critical branches are feeders of the critical chain, any delay on one of these would inevitably impact the critical chain. For this reason it is helpful to protect the critical chain tasks by inserting a *feeding buffer* in front of the non-critical branches. This also has a duration that is equal to the *pb* percentage of the non-critical branch in question.
- The list of the tasks in the project:

$$T = \{t_i\}_{i=1,\dots,n} = (t_1, t_2, \dots, t_n)$$

- The list of the task durations of the project:

$$D = \{d_i\}_{i=1,\dots,n} = (d_1, d_2, \dots, d_n)$$

- The list of the competencies necessary to carry out the tasks of the project:

$$TC = \{tc_{ij}\}_{i=1,\dots,n; j=1,\dots M}$$

where the matrix TC is made according to the same criteria as matrix K of the competencies previously introduced. For example, if $tc_{41} = 0$, this means that to carry out task t_4 the competency c_1 is not necessary.

- The list of the links that form the network of tasks that make up the project:

$$L = \{l_{ij}\}_{i,j=1,\ldots,n}$$

where L is a matrix (square, of n rows and n columns) with binary elements, defined in such a way that:

$l_{ij} = 1$ if the task t_i is a predecessor of the task t_j.
$l_{ij} = 0$ in all other cases.

3.7 Building the Critical Path

Once the variables described in the previous section have been introduced, and, in particular, once the competencies necessary for carrying out the tasks of the project and their logical network have been identified, we can move on to the actual scheduling. In this first phase we will identify the *critical path* and this is divided into two steps:

a. First we must identify the final task of the project. This must necessarily be one task (goal of the project). The property that characterizes the final task is that it has no successors. Thanks to this characteristic, it is fairly easy to identify; it is sufficient to go through the elements in the matrix of links L: due to the way this is defined, by reading the row i-th we can identify all the successors of task t_i; in the same way, by reading the column j-th, we will find all the predecessors of task t_j.

   Therefore, the final task of the project is identified in the only row of the matrix L where all the values are zero.

b. Once we have identified the final task, we can go on to identify the critical path. For this purpose, starting from the final task we go backwards through the project network, identifying all the possible ramifications, and for each one we calculate the length as a sum of the duration of the related tasks.

   By comparing the lengths of these various paths with each other, we identify the critical path as the longest path in terms of duration.

During the building of the critical path, we introduce a new square matrix $P = \{p_{hk}\}_{h,k=1,\ldots,n}$ the elements of which, initially all zero, are then partially substituted with suitable natural numbers during the scheduling.

The matrix P rearranges the network of the project according to criteria that are more convenient for subsequent calculations: once we have identified the final task t_i, its index "i" is inserted into the matrix P as the element p_{11}. As we will see, this is then replicated in an identical manner in the subsequent rows in the same column of matrix P, each time a ramification in the project network is found.

c. Supposing then that $p_{11} = i$, where t_i is the final task, we read the column i-th in L and, when we identify an element $l_{i_1 i} \neq 0$, we set $p_{12} = i_1$. This procedure is then iterated, meaning that we read the column i_1-th in L searching for predecessors of t_{i_1} and so on. When we get to a task that has no predecessors (i.e. a column in the matrix L made of null elements), the path $P_1 = \{p_{11}, p_{12}, \ldots, p_{1k_1}\}_{k_1 \leq n}$ has been completed (in practice it is always $k_1 < n$. Conversely, the project would be reduced to a single chain of n consecutive tasks).
d. When a task has more than one predecessor, a new branch is identified that will fill another row in the matrix P. If for example the final task t_i has another predecessor t_{i_2}, we set $p_{21} = p_{11}$ and $p_{22} = i_2$, and then we continue starting from this task, as in the previous step. We then identify a new path $P_2 = \{p_{21}, p_{22}, \ldots, p_{2k_2}\}_{k_2 < n}$ and so on.
e. At the end of this process, we will have obtained a certain number of paths: $P_1, P_2, \ldots, P_H$, each of which is made up of a vector of integer numbers that are the indices of the tasks of the related path. By calculating the sum of the durations of the tasks in each path and comparing the results, we can identify the longest one as the critical path of the project.

This is one of the most intricate and delicate phases of the algorithm. We tackle it by using a special recursive function that, at each step, determines either a predecessor of a certain task, or a ramification (when a task has more than one predecessor) i.e. a further sequence that may be the critical path.

In summary:

H is the number of different paths of the project. For every h such that $1 \leq h \leq H$, the path P_h is made up of k_h numbers that identify the number of tasks in the path P_h.

The whole iteration to build the critical path is carried out backwards, i.e. starting from the final task and going back through the project network to the starting tasks. Every time that, during this iteration, a new ramification is identified, the index of the different paths found is increased.

The new path that results from the bifurcation determines a new row in matrix P that will have the same elements as the path from which it has been generated up to the point of bifurcation.

If column j in matrix L has k elements that are not zero, then the task t_j has k predecessors and therefore this will generate $k - 1$ ramifications.

Once the critical path has been identified, it is placed into the calendar with the end date of the final task corresponding with the completion date of project E and, working backwards determines the start date of the first task in the critical path.

For example, let's imagine that we want to schedule a small project with six tasks. Figure 3.1 is a graphic representation of the project network.

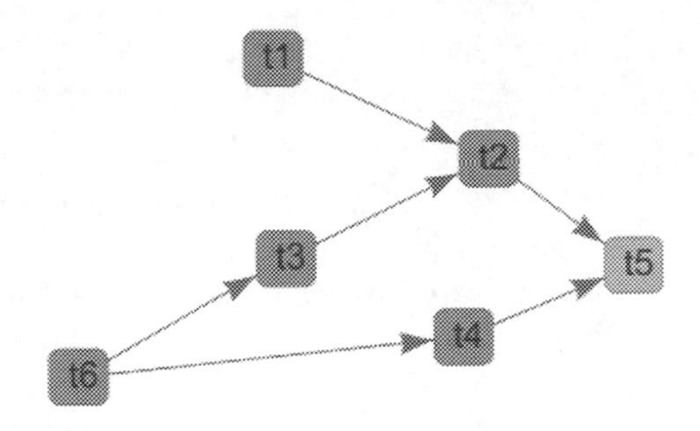

Fig. 3.1 Graphic representation of the project network

Starting from the network we can easily determine L, that contains the information about links among the tasks:

$$L = \begin{pmatrix} 0\,1\,0\,0\,0\,0 \\ 0\,0\,0\,0\,1\,0 \\ 0\,1\,0\,0\,0\,0 \\ 0\,0\,0\,0\,1\,0 \\ 0\,0\,0\,0\,0\,0 \\ 0\,0\,1\,1\,0\,0 \end{pmatrix}$$

The algorithm scans the rows first, ignoring the first 4 as they contain at least one element that is not zero. The fifth row, as it is null, determines the final task of the project: t_5.

After this, we set $p_{11} = 5$ and we analyze the fifth column, finding out that the second element is different from zero and therefore we can set $p_{12} = 2$. Therefore, for now $P_1 = \{5, 2\}$.

Moreover, as also $l_{45} \neq 0$, we set $p_{21} = p_{11} = 5$, $p_{22} = 4$ and therefore a branch has been identified: $P_2 = \{5, 4\}$.

To continue on the first path we must analyze column 2, where we discover that $l_{12} = l_{32} = 1$ and therefore we can set $p_{13} = 1$, hence $P_1 = \{5, 2, 1\}$ and, since there is another bifurcation: $p_{31} = p_{11} = 5$, $p_{32} = p_{12} = 2$, $p_{33} = 3$: $P_3 = \{5, 2, 3\}$.

The first column is null, therefore the first path is complete. Column 4 instead contains $l_{64} = 1$ and therefore $p_{23} = 6$ hence $P_2 = \{5, 4, 6\}$ and column 6 is null, which concludes path P_2.

Going back to P_3, column 3 is not null as $l_{63} = 1$, hence $p_{34} = 6$ and therefore $P_3 = \{5, 2, 3, 6\}$ so this path is also complete as column 6 is null.

Ultimately, in this project there are three paths that we can visualize explicitly by reading backwards the three paths found above:

$$\begin{aligned} P_1 &: \ t_1 \rightarrow t_2 \rightarrow t_5 \\ P_2 &: \ t_6 \rightarrow t_4 \rightarrow t_5 \\ P_3 &: \ t_6 \rightarrow t_3 \rightarrow t_2 \rightarrow t_5 \end{aligned}$$

And therefore matrix P is:

$$P = \begin{pmatrix} 5\,2\,1\,0\,0\,0 \\ 5\,4\,6\,0\,0\,0 \\ 5\,2\,3\,6\,0\,0 \\ 0\,0\,0\,0\,0\,0 \\ 0\,0\,0\,0\,0\,0 \\ 0\,0\,0\,0\,0\,0 \end{pmatrix}$$

Finally, we identify the critical path from the paths found as the one with the longest total duration.

3.8 Assigning Resources (Critical Chain)

Identifying the critical path is not the last stage in scheduling. The critical path in fact needs to be recalculated and completed with the assignation of resources to tasks to obtain the critical chain. This will provide us with the true duration of the project.

Identifying the critical chain starting from the critical path can be divided into four phases:

- First, the resources available in the system are assigned to the tasks on the critical path. During this operation there should not be any problem of resource contention as the critical path is a linear sequence of tasks that have already been placed into the calendar.

  When a resource r_i is assigned to a task, the starting and finishing dates determine an interval of time that is inserted as element $W_i(1)$ in the calendar matrix. If subsequently this resource is assigned to another task, a new interval of time $W_i(2)$, will be inserted into that resource's availability calendar.
- Next, we move on to assign resources to tasks on the non-critical branches. The tasks on a non-critical branch are analyzed one at a time, starting from the last one and working backwards. On the basis of the dates placed into the calendar for one of these tasks, the procedure is as follows:

  (1) We consider a competence that is associated with the task.
  (2) We filter the list of resources and select those that possess the competence from point 1.
  (3) For each of the resources identified in point 2, we compare their calendar availability with the dates of the task execution.
  (4) If one of these resources is not busy on those dates, it is assigned to the task.
  (5) If no resource with the relevant competence is available on those dates, then we are faced with a contention of resources.

- The third phase of scheduling is the resolution of resource contentions.

We tackle any resource contention due to an overlap in time (even partial) of different tasks that use the same resource. Resolving these conflicts may modify the critical path in such a way that certain choices become mandatory. The new sequence may even be quite different from the critical path. Necessarily, some of the tasks move to secondary branches and, in general, can be shifted backwards in time to avoid the overlaps that create resource contention.

In Fig. 3.2, tasks $T1, \ldots, T4$ in red are tasks on the critical path that have already been set in the calendar (every square block corresponds with the duration of one day) and to which resources with the appropriate competencies have been assigned to carry them out. Task $T5$, in blue, belongs instead to a secondary branch and, as can be seen in Fig. 3.2, partially overlaps with both task $T3$ and task $T2$.

If there are resources available on those dates with the necessary competencies to carry out task $T5$, and who are not busy with tasks $T2$ and $T3$, then the network will remain the way it is.

If, instead, task $T5$ conflicts with task $T3$ because there are not sufficient resources available for those dates to carry tasks T5 and T3 simultaneously, then task T5 is shifted back so that it no longer overlaps with task T3, as can be seen in Fig. 3.3.

With this move we have resolved the conflict between task T5 and T3 and, if T5 is not in conflict with T2 and T1 we can move on to the other tasks.

If, instead, T5 is still in conflict with T2, then rather than shifting it back further we have decided to insert it into the critical chain (every time we shift a task back it inevitably drags back all its predecessors, and this will very probably lengthen the duration of the project).

Therefore we move T2 and T1 back, in order to make space for T5, which in turn becomes a critical task, as shown in Fig. 3.4.

If, instead, task T5 is not in conflict with T3, but with T2, then in order to avoid too many reworkings, it is better to insert it into the critical chain, moving T1, T2 and T3 back, as shown in Fig. 3.5.

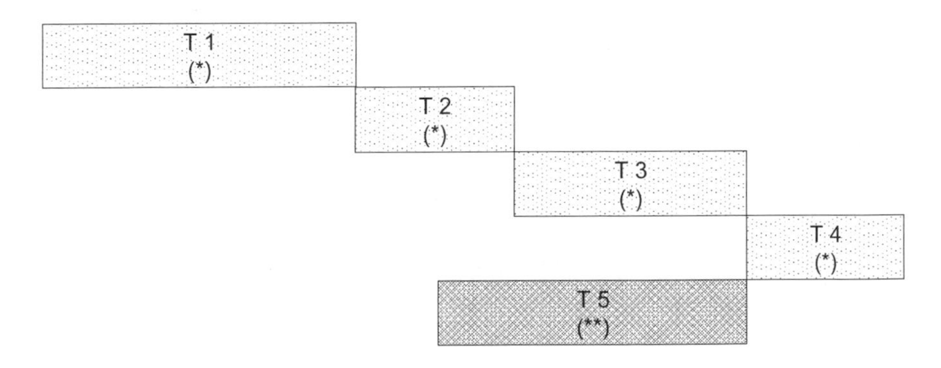

Fig. 3.2 Tasks on the critical path

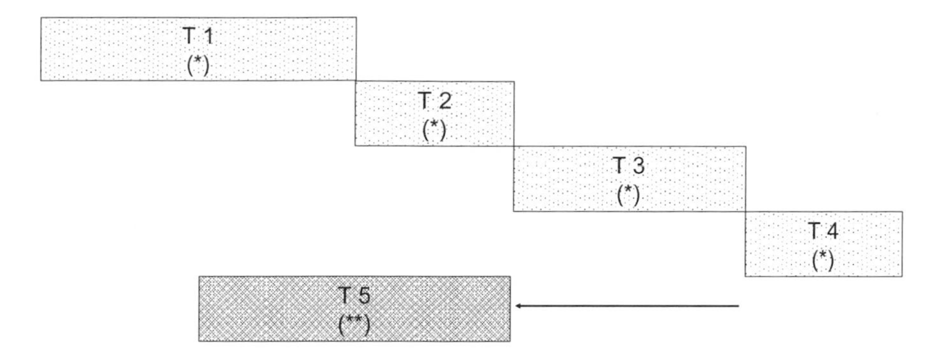

Fig. 3.3 Task 5 is shifted back

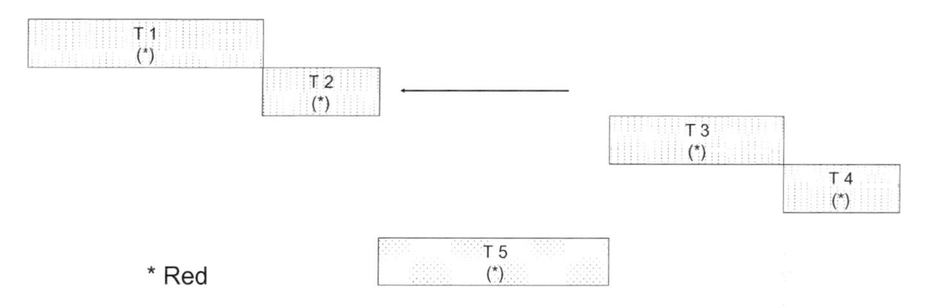

Fig. 3.4 Task 5 becomes a critical task

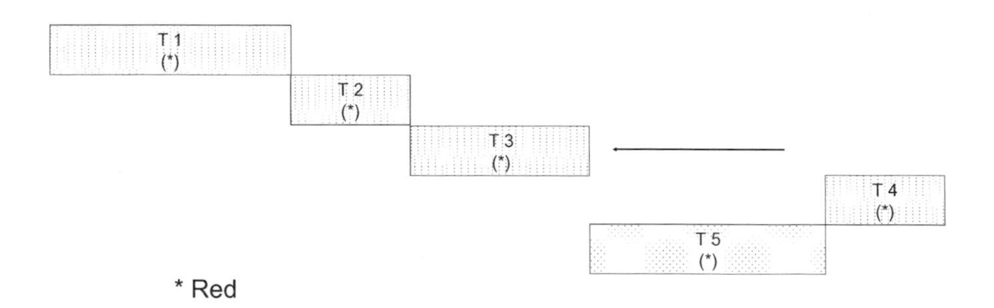

Fig. 3.5 Tasks 1, 2 and 3 are moved back

To illustrate the *modus operandi* of the algorithm, let's look again at the example of a project with six tasks from the previous section.

Let's suppose that there are three competencies in the system that are relevant. In Table 3.1 we list the properties of the six tasks in the project.

Table 3.1 Tasks and properties

Task	Duration	Competencies required	Resources with the competencies required
t_1	3	c_2 c_3	c_1: 2 resources
t_2	2	c_1 c_3	
t_3	3	c_2	c_2: 1 resource
t_4	4	c_3	
t_5	3	c_1 c_2	c_3: 2 resources
t_6	5	c_2 c_3	

On the basis of this information, given the network of the project, the three paths that we had identified have the following overall durations:

$$\begin{aligned} &P_1 : t_1 \rightarrow t_2 \rightarrow t_5 && \Rightarrow 8 \text{ days} \\ &P_2 : t_6 \rightarrow t_4 \rightarrow t_5 && \Rightarrow 12 \text{ days} \\ &P_3 : t_6 \rightarrow t_3 \rightarrow t_2 \rightarrow t_5 && \Rightarrow 13 \text{ days} \end{aligned}$$

Therefore the critical path is P_3.

Assuming we indicated a Friday as the completion date of the project, the project will be placed in the calendar, as in Fig. 3.6 (weekends are highlighted in grey as no activities are foreseen for these days. The scheduler shows them in the calendar but they do not, of course, add to the duration of the tasks).

Therefore, the tasks on the critical path are highlighted, as in Fig. 3.7.

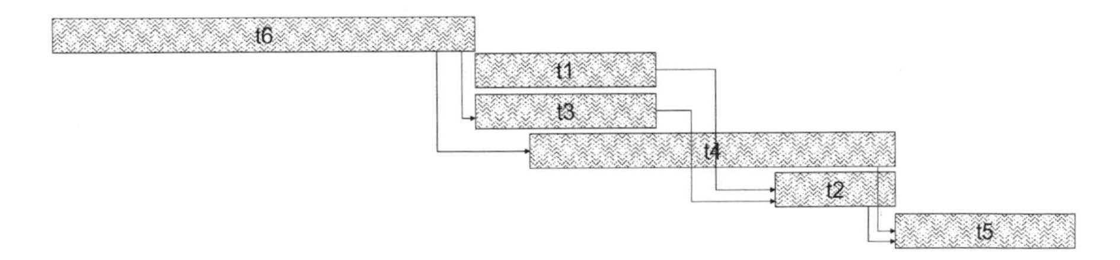

Fig. 3.6 The project is placed in the calendar

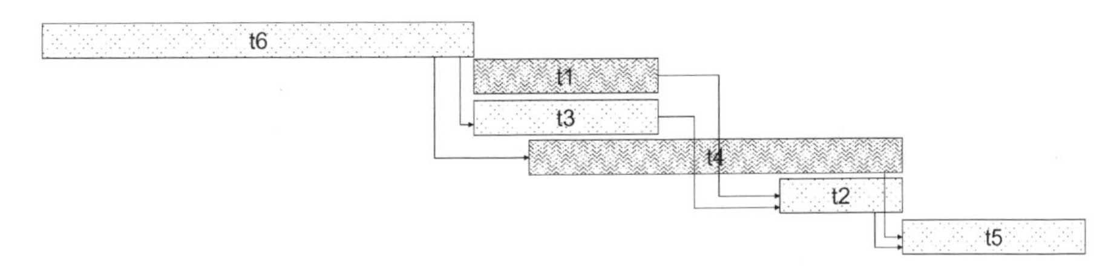

Fig. 3.7 The tasks on the critical path are highlighted

Regarding assigning the resources, there is no problem with the tasks on the critical path as they are consecutive and, therefore, we will find all the necessary resources to carry them out.

The non-critical task t_4, instead, partially overlaps with t_2 and t_3 on the critical path. Moreover, t_4 partially overlaps with t_1, which is instead on the other non-critical path.

There is no resource contention between t_4 and t_2 even if they both require competence c_3, because there are 2 resources available with this competence.

Moreover, there is no competence that is common to t_4 and t_3, so task t_4 can stay where it is.

Task t_1 is in conflict with task t_3 on the critical path because they both have in common the competence c_2, and only one resource has that competence. Therefore, task t_1 must be moved back to avoid overlapping with t_3.

After this operation, however, task t_1 will overlap with task t_6 and there will be resource contention with it, so t_1 can be inserted directly onto the critical chain, between task t_6 and t_3, or moved further back, to behind task t_6 so that it becomes the first task in the critical chain and therefore of the entire project (in this case the strategic choice for this last step does not make much of a difference because there are no tasks that are predecessors of t_6). In this way, the length of the critical chain changes from 13 to 16 days.

In this phase we proceed to establish the project buffer and feeding buffers size.

Let's suppose, in the previous example, to have attributed a protection $pb = 20\%$. Then, as the length of the critical chain is 16 days, the project buffer will be 4 days long (20% of 16 is 3.2 but we always round up to the greater integer number).

The project buffer lines up with the project completion date and the entire project is repositioned in the calendar accordingly.

Moreover, in this specific example, the only non-critical branch is made up of task t_4, that lasts 4 days. As 20% of 4 is 0.8 we assign a feeding buffer of 1 day to the task, and this will mean moving t_4 back by one day.

As this operation could create new resource contention, a further iteration is carried out in order to resolve any new conflicts.

Once this operation has been carried out, the project is repositioned in the calendar, this time taking into account all the buffers, as shown in Fig. 3.8.

The entire phase in which resources are assigned is organized in a rational way: if a certain competence is required to carry out a certain task, and if this competence is possessed by one or more resources in the system, it may be opportune to establish a selection criterion that takes into account the criticalities of the whole system.

The Ess3ntial algorithm organizes the resources according to a well-established order of criticality and it will first suggest resources that can be considered *less critical*, based on the following rules.

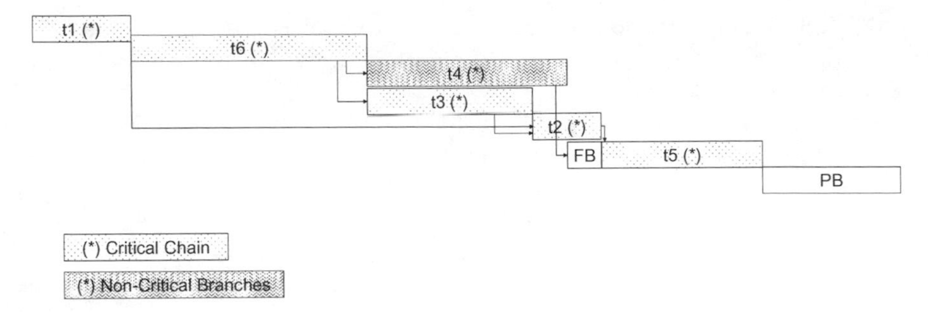

Fig. 3.8 Project is repositioned in calendar with buffers

The most critical resources tend to be those that have:

- more competencies;
- competencies that few other resources in the system possess (i.e. those that possess competencies that are *more rare* in the system);
- competencies that are highly requested within the specific project.

The algorithm manages these rules by means of suitable coefficients of criticality that are calculated during the scheduling phase.

3.9 Buffer Consumption

Once a project has been scheduled and launched, as soon as the start date of the first task has been reached, the commands that enable buffers to be monitored are automatically triggered.

Starting from the first critical task, it is possible to insert the days remaining until its completion.

Once the task is completed (an action that is certified by inserting zero days remaining in the relevant window), from the following day the next critical task can be updated. This is done until the project is completed, i.e. until zero days remaining until completion of the final task in the project is indicated.

Every day a value is manually inputted that allows the program to update the progress status of the project. Should manual input of this data be skipped, the program will automatically add a day of delay to the project execution and therefore a day of consumption of the project buffer is added.

The project progress status and buffer consumption status are shown in a graph. During the first two weeks of the project, buffer consumption is illustrated using the *three-zone method,* as in Fig. 3.9. This is an indicator in three colours that turns green when the buffer consumption is under 33% of its duration, yellow if the buffer consumption is between 33 and 66%, and red if the buffer consumption is above 66%.

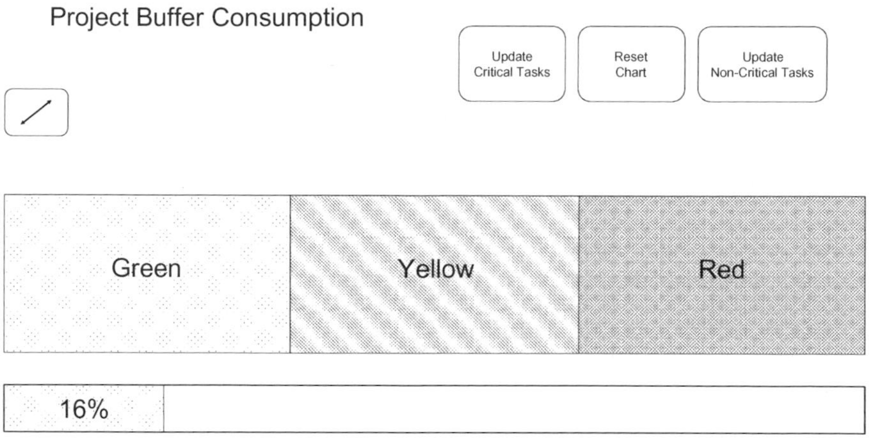

Project Buffer Consumption

Update Critical Tasks

Reset Chart

Update Non-Critical Tasks

Green

Yellow

Red

16%

Fig. 3.9 Three-zone method for buffer consumption

After the first two weeks, the modality of the monitoring of the project buffer consumption changes and it becomes a *control chart*; reading and interpreting these control charts requires knowledge of Statistical Process Control methods.

With reference to Fig. 3.10, the green line is made up of data that is added daily concerning execution of the critical tasks: an upward trend indicates a delay (i.e. an increase in consumption of the buffer), a horizontal line indicates daily completion that is carried out within the foreseen timeframe, a downward trend can be interpreted as time that has been caught up in reference to the data captured on the previous day.

The blue horizontal line indicates the length of the project buffer and it provides a reference point for the other data.

The black line represents the average consumption of the project buffer measured on the basis of data that is inputted daily.

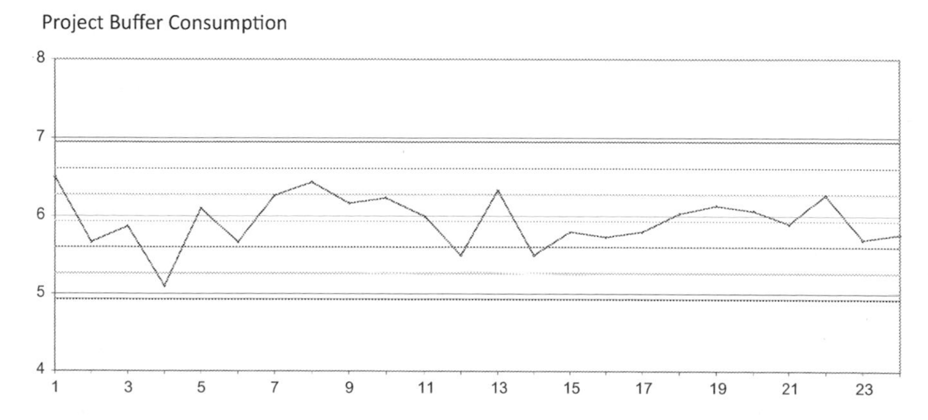

Fig. 3.10 Control chart of buffer consumption

The six lines positioned symmetrically above and below the average value are the upper and lower control limits (− 3sigma and + 3sigma,) and the limits at − 2sigma and + 2sigma and − 1sigma and + 1sigma that enable us to monitor the buffer consumption.

To be able to consider a process completely *in statistical control,* it should oscillate around the average value between the upper and lower control limits.

These control limits are calculated dynamically using the first 15 data points (i.e. during the "three zone" phase) after which they remain fixed.

The three-zone method is also used to monitor consumption of the feeding buffers but there is no shift to control charts because it would be superfluous (indeed, very rarely does a project have non-critical branches with feeding buffers that are longer than two weeks: if for example the protection percentage for the project is set at 20%, as is quite common, then that would mean having a non-critical branch that is about 75 days long!).

3.10 Multi-project Mode

The final characteristic of the Ess3ntial algorithm to consider is its multi-project mode.

Managing multiple projects simultaneously creates a considerable level of complexity. We must take into account not only the interdependencies within each project but also the those that may occur among the various projects and that must be carried out at the same time.

Ess3ntial enables the scheduling of any number of projects and also the scheduling of the same project as many times as may be desired. The resources assigned to the tasks are not booked in their calendars until a project is actually launched.

Once a project has been launched, the resources assigned to it may then not be available for other projects that must be carried out more or less in the same time period. Therefore, a project that has been scheduled previously may no longer be suitable to be launched due to the unavailability it creates for its assigned resources.

In this case, it is possible to reschedule the project, in the hope that there are enough alternative resources so that it can be carried out more or less on the same dates or, alternatively, the project can be scheduled from scratch with a modified completion date.

Every time a project is scheduled in Ess3ntial in a period when other projects are due to be carried out, the algorithm always takes into account the resources that are not available and it resolves any resource contention by applying the same measures that we covered previously.

Should some resource contention prove unresolvable, either within a single project or multi-project mode, Ess3ntial communicates the error and indicates which task and which competence is encountering the problem.

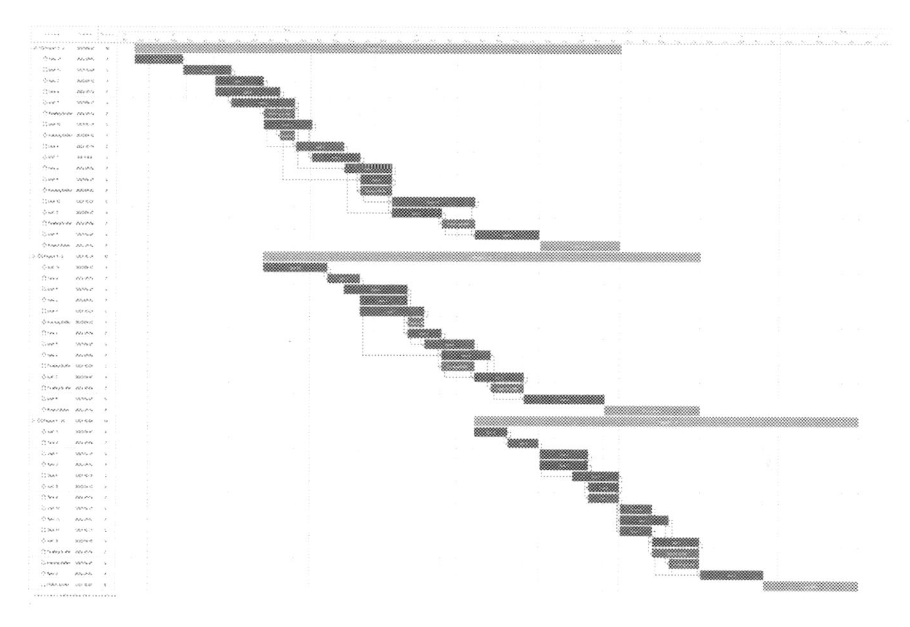

Fig. 3.11 Multi-project display in Ess3ntial

When this happens, there are only two options:

(1) Schedule the project in another timeframe with a later completion date;
(2) In case there is a delivery date that cannot be postponed, hire new resources (or use external resources for short periods in order to get through the emergency).

In Fig. 3.11, you can see the multi-project display in Ess3ntial, showing the simultaneous scheduling of three small projects that overlap in time with no resource contention and using the same pool of competencies.

3.11 The Success of the Organization

The orchestration of competence-based activities geared towards an identified goal is the very definition of "Project". The coordination of all these projects is made far easier by having a larger pool of competencies to draw from; the number of projects that can be handled in a given period increases and all the talents available can contribute fully. What we achieve by synchronizing these available competencies across projects is to maximize what existing competencies can contribute to the success of the organization.

Chapter 4
From Silos to Network—The Intended Consequences

Abstract In this chapter we make some considerations about measuring and accounting for performance, growing the system sustainably, hiring and developing individuals in the organization, what governance means and how it can be improved, and how an important, systemic shift can be achieved operationally. We stress the importance of continuous learning in the shift from silos to network.

Keywords Throughput accounting · Governance · Continuous learning · Thinking processes · Human resources

We hope that the reasoning presented in the previous chapters has illustrated *why* a shift in organizational design is needed to embrace Complexity and *how* it can be done through the Network of Projects.

However, none of this reasoning or knowledge can be of any use to an organization unless we address those factors that enable the system to be sustainable over time. The work of *managing* Complexity calls for a complete, comprehensive overhaul of all the prevailing assumptions that still today guide most organizations. The most important of these assumptions affect the ability to:

(a) measure and account for performance;
(b) feed a pipeline of prospects sensitive to a value-based business proposition;
(c) hire and develop the individuals within the organization.

In this chapter we would like to make some considerations about these aspects of management and how an important, systemic shift can be achieved operationally.

4.1 Accounting and Speed

Forty years ago, in a paper presented to a gathering of CPAs, Dr. Goldratt declared that Cost Accounting and the intricacies of GAAP accounting were the *Enemy #1 of Productivity*. In the now legendary "P&Q" exercise (1) Goldratt exemplified live to an astonished audience the inherent impossibility for conventional accounting to

D. Lepore et al., *From Silos to Network: A New Kind of Science for Management*, SpringerBriefs in Complexity, https://doi.org/10.1007/978-3-031-40228-9_4

grasp the essence of a constrained system. The P&Q exercise echoes similar experiments conducted by Deming on variation known as the "Funnel experiment" and the "Red Bead experiment" (2). Both of these experiments were aimed at unveiling to leaders and managers that the complexity generated by interactions and interdependencies generates emergent properties that cannot be explained simply in terms of their components, rather we must consider *how* these components interact and interdepend.

This should not be a surprise. GAAP accounting rests on the "Double Entry" method, developed in Italy by a Franciscan friar long before Galilei and Newton gave us Calculus and the ability to understand "motion". Indeed, in accounting there is no "motion picture", there are only snapshots. It is small wonder that finance people on Wall Street constantly sidestep GAAP with all their "speed-based" financial gimmickry to accelerate so-called value creation, with the ensuing devastation that is evident to us all.

An appropriate form of "Accounting" for the new economics underscored by Complexity, simply speaking, is not Accounting at all. Its goal is not to ascertain *the nature of costs* but to unveil all the *possibilities to create value* for the market through operations. Dr. Goldratt created a suitable measurement system with a way of highlighting possibilities for organizations to improve drastically their performance: he called it *Throughput Accounting*. Throughput Accounting is a *systemic measurement approach* based on speed; it is a *derivative* measure that looks at cash generation per unit of time. Throughput Accounting must not be "reconciled" with GAAP Accounting, nor is there any need to do so because the two systems have different goals. What we can say here is that, over time, if a company pursues Throughput (speed of cash generation) this will accommodate for any concerns induced by GAAP Accounting that are related to Profit and Loss; the Balance sheet is almost irrelevant for Throughput Accounting purposes.

4.2 The Cost of Cost

Any form of officially recognized "Accounting" is based on the concept of *cost allocation*. This concept rests on the belief that when a company purchases any form of raw material and transforms into a good or a service, every stage of that transformation creates a further, hidden cost. Consequently, these hidden costs must be allocated to the raw material in order to ensure that an appropriate sales price can be attached to the finished good or service to produce a "profit margin".

Indeed, the guild of cost accountants believe that products have a "cost" and their job is to find out what it is in order to guide organizations to determine a sales price for each product that "ensures" profitability. As a result, companies often turn down potential orders, forfeiting cash generation and market share, because through the lens of cost accounting the order is not "profitable" enough.

The devastation driven by cost allocation would merit an entire treatise; a reminder that the sales price of a good or service is set by market perception, not by arcane calculations done by accountants, would simply not suffice. The whole concept of "product cost" is one of those *cognitive constraints* that truly debase the ability of organizations to thrive.

Goldratt never wrote anything sufficiently comprehensive to change accountants' minds about the scope of their profession (accountants sincerely believe they can "unlock value" for organizations through the number manipulations that GAAP allows), nor did he pay much attention to the organizational implications of Throughput Accounting. He left this job to those willing accountants (often with a tenuous grasp of Complexity) who have produced over the years a number of publications that served the purpose of stirring some curiosity. Not without some hesitation, we can recommend Debra Smith's 'The Measurement Nightmare' (3) for a primer on the subject. For a much shorter and informal introduction to Throughput Accounting, see our books 'Sechel' and 'Quality, Involvement, Flow' (4).

4.3 What's in a Name?

The very name Throughput Accounting can create some confusion, precisely because it contains the word "accounting". But what's in a name? They can be mysterious at times and can even disguise paradigmatic changes in the way we perceive reality around us. A name can be a paradox because it can simultaneously reveal and disguise the nature of something. We can thank Rabbi Chaim Miller for pointing this out in his comments at the beginning of the section of the Torah known as Exodus, in Hebrew called 'Names' (*Shemot*). Based on a Sichos (lesson) given by Rabbi Menachem Mendel Schneerson, the Lubavitcher Rebbe, Miller reminds us that a person's name reveals nothing as it is common to many people, but at the same time, a person feels that their name reveals their uniqueness. Without having any particular biblical pretensions, we think that the name Throughput Accounting is one of these paradoxes. (5) Throughput Accounting, arguably a lame and misleading name, reveals the essence and can be the engine of a *new economics* (as advocated by Dr. Deming) based on a win–win, whole system optimization.

4.4 Exponential Growth

In a complex and increasingly global world, we need an entirely different lens in order to grow and thrive. In order to present a systemic solution for growth that did not rely on the artificial manipulation of accounting, Dr. Goldratt published 'It's Not Luck' in 1994 (6). In this business novel, he addresses Marketing, Sales and Distribution—what he called *non Physical Constraints*. For the first time, he formalized some of the thinking processes (7) necessary to systematically unveil hidden, unverbalized mental models (he calls them *assumptions*) that stifle organizations in addressing situations of *external constraint*. This External Constraint is where the limitation is dictated by the unwillingness of the market to buy all the available capacity the system could potentially deliver.

A few years later, Goldratt launched the *Viable Vision* program. Through this program Goldratt claimed that in four years it is possible to take (almost) every organization to a net profit in year four that is equal to the total current sales in year one. You heard us right: if "year one" sales is 100 and profit is 10, then in "year four" profit becomes 100. Viable Vision is by all means "viable" but it requires a critical shift in thinking in order to grasp the complexity generated by the role that the constraint of the organization plays in value creation.

4.5 Why the External Constraint Matters so Much

In this book, our intention is to offer a "viable" systemic organizational redesign based on a network of constraint-based projects scheduled at finite capacity. However, we feel that a few words are in order to highlight the full extent of what it takes to transform from Silos to Network.

Constraints, *whether physical or not*, are not an objective reality. They are the results of our intrinsic inability to perceive the entirety of cause-effect relationships that shape reality as we experience it. This inherent limitation of our learning apparatus is made up of a collection of consolidated beliefs, often generated by real life experiences. Over the years we have come to call this limitation our *cognitive constraint*. Accordingly, any improvement, let alone transformation, can only start with addressing such a constraint.

Designing an organization with a constraint in mind, physical or otherwise, entails challenging deep-seated and long standing views and practices that simply make no sense. For example, it is common practice to offer discounts for higher volumes of the same product; higher volumes imply bigger utilization of the same resources. This is something that accountants call "efficiency" and it is a concept they cherish. The problem is that often among these resources there is the "constraining" resource. By offering a discount, we are basically diminishing the amount of dollars per hour that we generate through that constraining resource. As the constraint dictates the pace with which we generate throughput, this means diminishing the profitability of the entire system regardless of the improved efficiency of the resources. For large volumes of the same product we should actually charge more!

Even more dramatically: when we strive to increase the efficiency of our resources (percentage of utilization time in respect with time available) all we obtain is a pile of inventory throughout the system waiting for the constraint *and* a progressive increase in lead time because everybody wants to keep busy doing things that are not required so that they can look more "efficient" (TOC calls this "stealing"). To add insult to injury, the balance sheet will look great because all that WIP will be "valued" financially while gaining dust on the shelves.

Probably, the most dramatic shift in learning that an organization needs to undertake in moving from silos to network is in the way it approaches the market. Goldratt's *External Constraint* solution is a systemic approach to Marketing and Sales; it is the direct result of managing the constraint and, more in general, the network of projects underpinning it. Some of the authors of this book have been practicing this approach for many years with perpetual amazement at the results achieved.

Just as we describe in Chapter One in regard with Project Management, External Constraint is originated by the simultaneous need that a company has to sell *all* its capacity *and* to see recognized the value that it brings with its products. (For the inherent conflict of External Constraint, see Appendix D.)

To address both of these needs it is necessary to embrace a paradigm of togetherness with the network of suppliers and customers (often suppliers of suppliers and customers of customers) that is loosely based on the understanding that:

- It is not a single organization that competes, rather the Value Network it is part of;
- The selling price is determined by the market's perception of value *not* product cost; the floor of this price is dictated by the Totally Variable Costs (e.g. raw material + shipping) of the products and amount of time the constraint is utilized;
- Each market segment (and each supplier) has its own predicaments and they must be clearly understood in order to maximize value perception;
- Nobody in the Value Network makes any money sustainably until the end user is happy to pay for the product;
- It is not about selling, it is about creating a partnership-based buying environment;
- The Economics of the Value Network is different from the Economics of the single node in the Network.

External Constraint is a quintessentially systemic exercise as it looks at the marketing and sales effort of a company as activities that are inherently connected with the inner workings of the organization, not a separate department, hence generating complexity.

Part of this complexity is the hiring, onboarding, deployment and development of individuals in the organization. Traditionally, this is the work of a department, the HR department. Once again, any temptation to see company activities in a vertical way and as separate from other activities in the organization not only suboptimizes the efforts of everyone in the company, it perpetuates a fragmented vision of what an organization really is. "HR" instead, embodies a series of ongoing activities company wide and this requires a shift in our understanding of how to manage humans and their competencies effectively towards a common goal.

4.6 Putting Humans and Resources Back in HR

In traditional organizations, HR is considered as a "department" that takes care of all the issues connected with legal compliance and employees (contracts, onboarding, offboarding, etc.). It may also be responsible for the training and development of human resources and, to some extent, individual performance evaluation.

When we look at an organization as a whole system, it makes little sense to have an "HR department". Firstly, because there are no departments in a systemic organization oriented towards a common goal. There are *interdependencies* that happen through processes and projects. The people who interact through these processes and projects do so because they have the competencies that are required.

All of these interdependencies, whether they are through written or spoken communication, happen through language. The role of HR, then, is to *support the management of interdependencies* among people who work on processes and projects that are structured and commonly understood.

In a systemic organization, there are no artificial barriers or silos and the emphasis is on speed of flow. Problems can occur that interrupt the flow of human interactions. Precisely because they are human, these interactions are never purely rational. HR has a special role in supporting and strengthening a process-based way of managing the organization concerning those aspects of people's interactions that are not strictly rational and where emotions are involved.

The Theory of Constraints provides powerful and reliable Thinking Processes to guide and manage interactions that can harness people's emotions in a positive direction towards the common goal. First introduced in Goldratt's novel 'It's Not Luck', we have described these Thinking Processes in some detail in previous publications (8).

4.7 Benefits of Using the Thinking Processes

We all struggle with going beyond linear thinking to embrace solutions that may seem counterintuitive. To shift towards a systemic way of interpreting and interacting with reality requires continuous effort.

The Thinking Processes from the Theory of Constraints (TOC) represent a linguistic framework that amplifies and fortifies our cognition. They allow us to identify, manage and overcome the obsolete mental models that prevent us from conceiving the broader reality of interdependencies in which we are all immersed. They help us paint a new epistemological framework that is suitable for complexity.

The Thinking Processes aid and strengthen our ability to harness the power of our emotions and support the change process, more specifically:

(1) What to change (intuition)
(2) What to change to (analysis)
(3) How to make the change happen (implementation)

The many benefits include:

- visualizing the complex, highly nonlinear network of cause–effect relationships that make up reality, as we perceive it;
- mapping the "conversations" that define our cognitive horizon;
- conceiving of new ideas (intuition) and their full development through analysis (understanding);
- carrying out the implementation of the fully analyzed idea (knowledge);
- conceiving breakthroughs that did not previously seem possible;
- reducing variation in our thought processes by focusing our mental efforts toward a goal;
- reducing variation in the way people communicate in an organization by providing a common language;
- reinforcing and engendering the empathy required for collaborative work;
- helping people to manage the interdependency of intellect and emotion in the change process.

4.8 HR and Business

Many people would have difficulty in associating HR with the business side of a company. This is the result of a fragmented understanding of an organization. HR, like every other part of the organization, is there to contribute to Throughput generation. To achieve this, HR must have a positive impact on the way people operate within the system. One aspect of this is helping with speed of flow in interactions, as described in the previous section. HR also has an important role in mapping out the competencies the organization needs to reach its goal in a continuously improving way. HR contributes to this specifically through the ongoing upgrade and intake of necessary competencies as the company grows.

When HR participates in meetings, they can very fruitfully summarize what goes on, and keep notes to identify interactions that can be framed linguistically as conflicts, assumptions, negative implications and action items that require planning and execution. These *speech acts* can be captured effectively using the Thinking Processes from the Theory of Constraints as mentioned previously. HR can also facilitate the interactions by keeping them on track and focused when conversations drift off the agreed subject matter. The focus is always throughput generation for the company as that is how everyone involved benefits, all throughout the value network. By continuously making use of the Thinking Processes, HR contributes directly to the business by ensuring speed of flow in people's interactions and therefore speed of throughput generation. Everybody wins.

4.9 HR and Psychology

In a systemic organization, HR's focus is not individual performance coaching but continuously enhancing interactions within the system to accelerate flow. Performance reviews are not appropriate in a systemic organization as it is the system, not the individual, that creates the parameters and dynamics for what can be achieved. If individuals are having specific problems psychologically or cognitively then, of course, outside support can be engaged. What HR *can* act upon are the interdependencies and interactions among people. Without smooth interactions the company will have more difficulty in achieving its goal of throughput generation. The Thinking Processes help people adopt an effective mental stance for working systemically and this, in turn, increases the collective intelligence within the company. In this sense, the concern is not psychology but rather philosophy of the mind—how we think and how we can think better, and epistemology—how we know what we know and the role of mental models and assumptions in the way we work towards our common goal.

4.10 A Word About Motivation

Some traditional HR areas of focus include "motivation". As Deming has pointed out, if management stopped demotivating their employees then they wouldn't have to worry so much about motivating them.

In a systemic organization, the emphasis is on designing *correct interdependencies* that can stimulate people's natural, *intrinsic* motivation to work and where their behavior is not influenced by misguided incentive systems (*extrinsic motivation*).

4.11 Continuous Learning

External reality continues to change at a hectic pace and any organization must still guarantee that it is working properly and achieving its goal. This means ensuring a mechanism is in place that allows it to generate and update the knowledge needed in order to manage the system.

A systemic organization is one where continuous improvement is at its heart. To achieve this goal—to improve—it must embark on the process and the mindset of continuous learning. HR can contribute in a significant way to the management of this learning.

4.12 What Affects Our Ability to Learn

Learning to understand and manage a company as a system is, for most organizations, a transformational change. For humans to translate what they know into coherent behaviors, we have to understand the fundamental forces that act upon them and make them who they are. This is because these forces do influence the cognition process and limit or enhance the ability to learn.

As humans we experience simultaneously a fundamental need for restraint, or *control* and a drive or desire to *go beyond* our current state and see ourselves projected into a different and greater dimension of existence. An expression of these drives can be found in two, seemingly conflicting desires that humans have. These are the desires to:

- Belong and to bond (be part of a community).
- Develop our unique, distinct identity (be an individual).

Finding a way to harmonize these desires is central for a thriving organization. How can the interests of the community (organization) and the individual (employee) all be fulfilled and served? As Dr. Deming repeatedly pointed out, an individual's performance within an organization is the result of what the individual is capable of *plus* all the interactions and interdependencies in which they are immersed. As a result of these interactions, Dr. Deming rightly claims that assessing individual performance in an organization is both dangerous and useless.

An organization has to build the right kind of *interdependencies* so people do not feel imprisoned in their work and there is *intrinsic meaning* in what they do. This means neither dependence nor independence, but interdependence. In this way, the individual can understand that by contributing with intelligence, passion and adherence to the company goals, they derive more perceived benefit than they would by working alone (being independent).

How can the company leverage people's natural desire to bond and belong? By providing opportunities to participate in something from which the benefit they derive is greater than the effort they put in. For an organization to allow this meaningfulness to exist in the workplace it must start from a vision. It must then create the right interdependencies, after which it is up to the individual to take part and become one with the organization knowing that their own personal life will be enhanced. This is precisely what the multi-project environment can offer: people contribute what they are capable of and the system leverages this. By enabling people to do what they are good at there are immediate good results that reflect back on the worker in a plurality of positive ways.

What is the real difficulty that we face in wanting to create a true learning organization that dismantles the functional structure and replaces it with the far more suitable network of projects? It is not connected with lack of knowledge of how to do it, nor with the lack of technologies to support it. The real issue is the mental barrier, or, as we have come to call it, the *cognitive constraint* that prevents individuals and organizations from working together as a system for a common goal. We cannot afford to be pessimistic about the possibility to elevate that cognitive constraint. It requires continuous attention and effort to enable people to shift from the paradigm of traditional, hierarchical/functional organizations towards an organization design that is fit for our times.

4.13 Do We Even Need HR at All?

Once again, the issue of how we name something is relevant as it affects our understanding and the assumptions we make. Perhaps it is time to replace the term "HR" with something more appropriate for our age of complexity. After all, what we are seeking to manage is not so much "human resources" as opposed to non-human resources. We are faced with a continuous effort to manage the most effective and satisfying interactions possible with the competencies available. This is a responsibility that is distributed throughout the organization. There will always be the need for a person or persons to be a reference point for this work but they are certainly not a department.

4.14 A Brief Note on Governance and How to Improve It

Industry needs Investors (the Capital) to grow and Capital needs Industry to be effective in deploying it so they can get their returns. The interface between Investors and CEOs (those who guide Industries) are Boards, who have the mandate to protect the Capital from any misuse, fraud, embezzlement, malpractice, illegalities, etc. that could be generated by the work of the Industry that benefits from that Capital. Also, Board members are valued for their ability to provide real life advice to CEOs based on their expertise.

In addition to their gut feeling and past experience, Board members rely on tools to exercise their duties that are based on GAAP accounting (and the letter of the Law). From Enron to FTX we have repeatedly witnessed the spectacular failure of their efforts to provide any insight or foresight, let alone any guidance, to CEOs and their inability to protect the Investors they represent (but in the meantime they have created plenty of business for law firms).

Boards normally meet twice a year and are fed numbers prepared by CFOs and Accountants that essentially tell whatever story the CEO wants to portray. They have *zero* possibility to discern what the real business situation is and are just happy to hear about the glorious efforts of the Management team and the inevitable great results that will happen in the future. They can only ask vague, generic questions that any skilled CEO can answer pretty much any way he/she likes. Good dinners in nice locations help to smooth the edges of these meetings. When things start going sideways, "activist investors" step in with all the ensuing chaos.

Do we really need Boards? Of course we do. The question is: what should they do? How could they realistically assuage the concerns of the investors and provide real value to an enterprise?

Let's stick to the basics.

Any organization, let's say for the sake of simplicity a "for Profit" organization, has the goal of generating cash, now and in the future. *If* an organization accepts the fact that they generate cash through projects *and* that projects must be managed in the way we tried to explain earlier, *then* the only realistic, value generating role for a Board is to ask CEOs questions regarding:

- the state of the buffer for existing projects (so they *really* understand what short term revenues will be);
- the state of the pipeline of projects (medium term revenues);
- what new initiatives are about to be launched (so they know where the CEO is taking the organization (long term revenues);
- The financial implications of the above.

Board meetings should then become short, frequent conversations with Top Management about how the system is evolving as result of the projects chosen. A Board well versed in the inner workings of a transparent organization (Buffers do not lie) can then truly become the guarantor to the Capital Investors of the way money is being deployed.

Notes

1. Goldratt (1990).
2. Deming describes both the Funnel experiment and the Red Bead experiment in Deming (1993).
3. Smith (1999).
4. See in particular Chapter 11 in Lepore (2010) and chapter 12.6.2 in Lepore et al. (2017).
5. Rabbi Chaim Miller refers to a lesson given by Schneerson (2002).
6. Goldratt (1994).
7. For detailed descriptions of the Thinking Processes from the Theory of Constraints see our books Lepore and Cohen (1999); Lepore (2010); Lepore et al. (2016, 2017).
8. See Lepore and Cohen (1999); Lepore (2010); Lepore et al. (2016, 2017); Lepore (2019).

References

W.E. Deming, *The New Economics for Industry, Government* (Education. Massachusetts Institute of Technology Center for Advanced Engineering Study, Cambridge, MA, 1993)

E.M. Goldratt, *The Haystack Syndrome: Sifting Information from the Data Ocean* (North River Press, Great Barrington, MA, 1990)

E.M. Goldratt, *It's Not Luck* (North River Press, Great Barrington, MA, 1994)

D. Lepore, *Sechel: Logic, Language and Tools to Manage Any Organization as a Network* (Intelligent Management Inc., Toronto, 2010)

D. Lepore, *Moving the Chains: An Operational Solution for Embracing Complexity in the Digital Age* (Business Expert Press, New York, 2019)

D. Lepore, O. Cohen, *Deming and Goldratt: The Decalogue* (North River Press, Great Barrington, MA, 1999)

D. Lepore, A. Montgomery, G. Siepe, Managing complexity in organizations through a systemic network of projects, in *Applications of Systems Thinking and Soft Operations Research in Managing Complexity*. ed. by A. Masys (Springer International Publishing, Switzerland, 2016), pp.35–69

D. Lepore, A. Montgomery, G. Siepe, *Quality, Involvement, Flow: The Systemic Organization* (CRC Press, New York, 2017)

R.M.M. Schneerson, the Lubavitcher Rebbe (Sichas Shabbos Parshah Shemos 5746) at the beginning of the Gutnick Edition of Shemos, in *The Book of Exodus Kol Menachem*. ed. by C. Miller (New York, 2002)

D. Smith, *The Measurement Nightmare: How the Theory of Constraints Can Resolve Conflicting Strategies, Policies, and Measures* (CRC Press, New York, 1999)

Appendix A
The Add Protection Don't Add Protection Conflict

The Conflict Cloud is a Thinking Process from the Theory of Constraints. It is a highly effective way to frame a complex situation by thinking of it in terms of conflicting positions, legitimate needs that provoke those conflicting positions, and a common goal. We can then surface all the assumptions (mental models) that keep us stuck in any given conflict or situation of blockage. Once we have identified all these elements, we have a way to allow a win–win solution to emerge. This Thinking Process can be used at a variety of levels, from solving day-to-day problems to analyzing and solving highly complex issues.

The diagram below explains the contents of a Conflict Cloud and how to read it once it has been built (Fig. A.1).

D. Lepore et al., *From Silos to Network: A New Kind of Science for Management*, SpringerBriefs in Complexity, https://doi.org/10.1007/978-3-031-40228-9

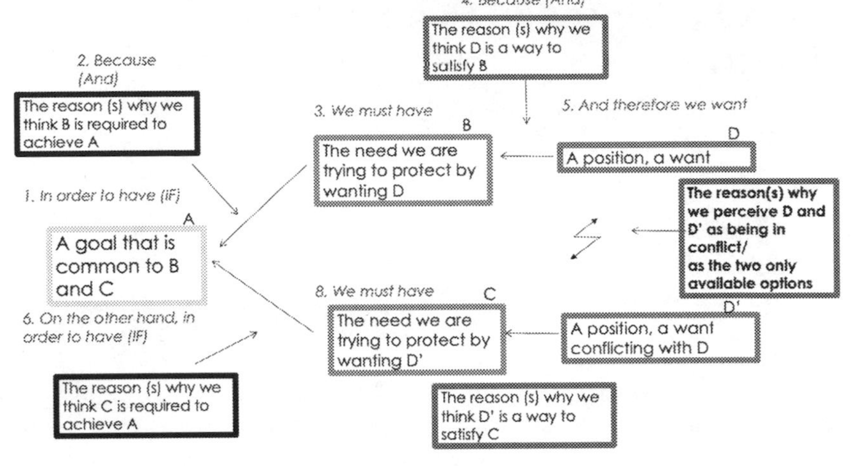

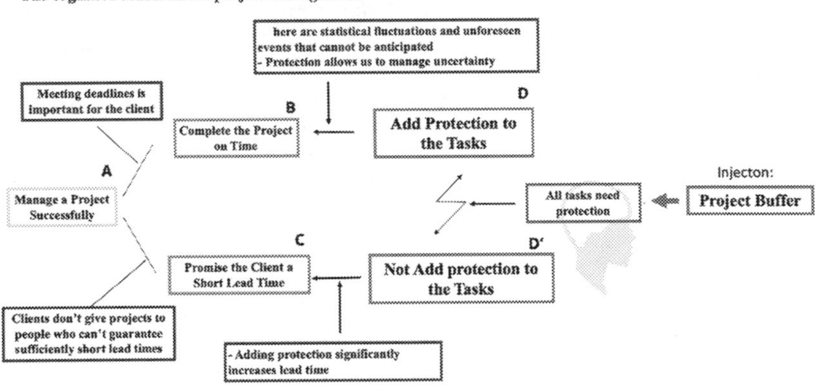

Fig. A.1 **a** Template of a conflict cloud. **b** The add protection don't add protection conflict

Appendix B
The Start Early Start Late Conflict

See Fig. B.1.

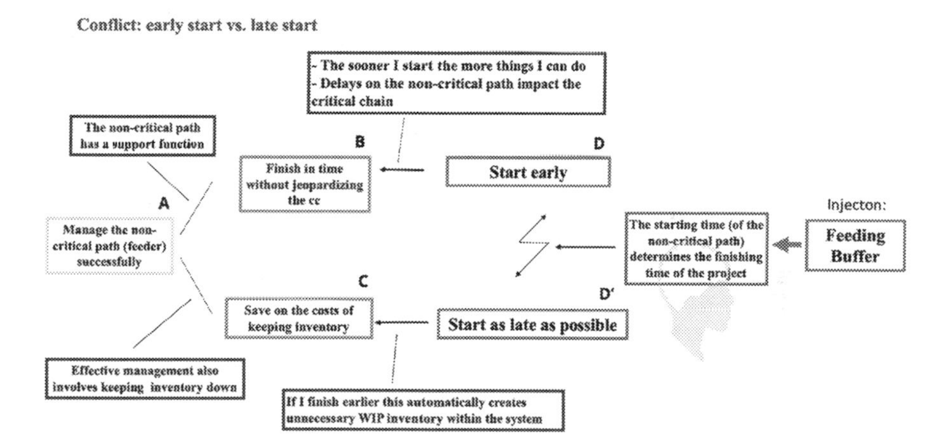

Fig. B.1 The start early start late conflict

D. Lepore et al., *From Silos to Network: A New Kind of Science for Management*, SpringerBriefs in Complexity, https://doi.org/10.1007/978-3-031-40228-9

Appendix C
The Intervene Don't Intervene Conflict

See Fig. C.1.

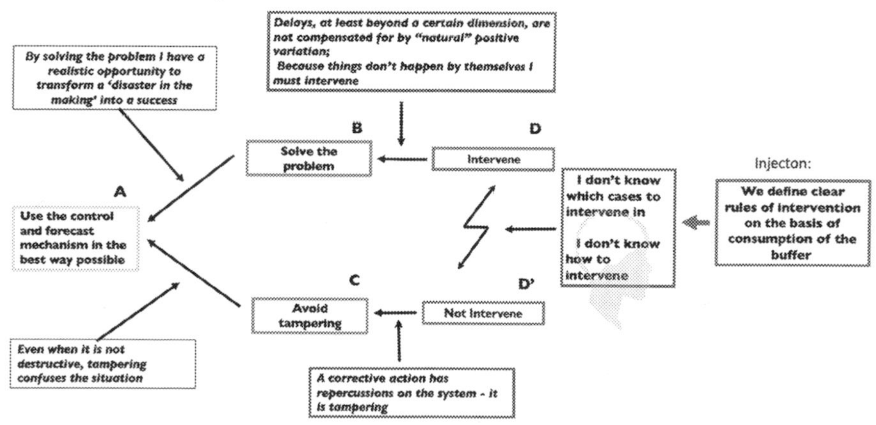

Fig. C.1 The intervene don't intervene conflict

D. Lepore et al., *From Silos to Network: A New Kind of Science for Management*, SpringerBriefs in Complexity, https://doi.org/10.1007/978-3-031-40228-9

Appendix D
The External Constraint Conflict

See Fig. D.1.

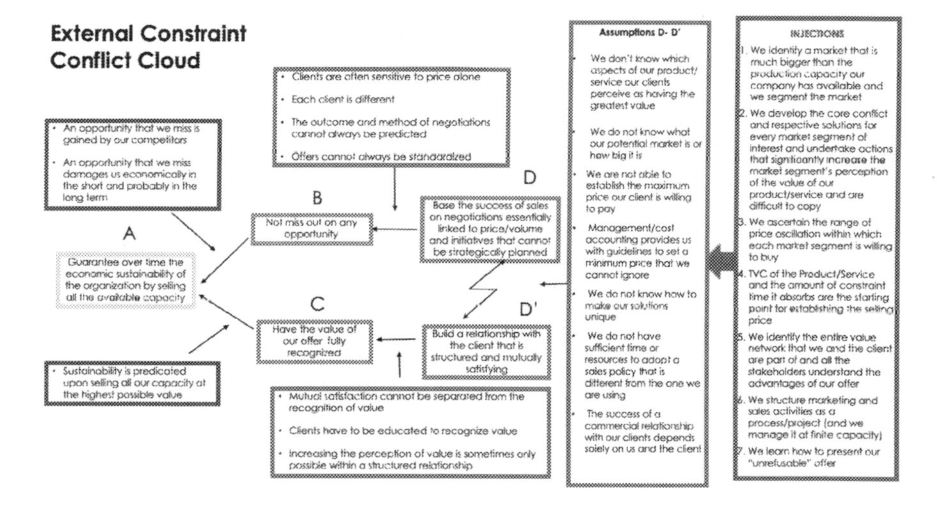

Fig. D.1 The external constraint conflict

D. Lepore et al., *From Silos to Network: A New Kind of Science for Management*, SpringerBriefs in Complexity, https://doi.org/10.1007/978-3-031-40228-9

Bibliography

The following bibliography does not represent an exhaustive list of the resources that have been studied as part of the development of the work described in our book. This list does, however, provide a solid basis for an approach to all the aspects of systemic management, as proposed in The Decalogue methodology and the Network of Projects organizational design.

A. Barabàsi, A. Reka, Emergence of scaling in random networks. Science **286**(5439), 509–512 (1999)

A. Barabàsi, *Linked: The New Science of Networks* (Perseus Publishing, Cambridge, MA, 2002)

F. Capra, *The Web of Life: A New Scientific Understanding of Living Systems* (Anchor Books, New York, 1996)

F. Capra, P.L. Luisi, *The Systems View of Life: A Unifying Vision* (Cambridge University Press, Cambridge, 2014)

R. Cialdini, *Influence* (Harper Business, New York, 2006)

T. Corbett, *Throughput Accounting* (North River Press, Great Barrington, MA, 1998)

W.E. Deming, *Out of the Crisis* (Massachusetts Institute of Technology Center for Advanced Engineering Study, Cambridge, MA, 1986)

W.E. Deming, *The New Economics for Industry, Government, Education* (Massachusetts Institute of Technology Center for Advanced Engineering Study, Cambridge, MA, 1993)

E.M. Goldratt, *What is this Thing Called the Theory of Constraints and How Should It Be Implemented?* (North River Press, Great Barrington, MA, 1990a)

E.M. Goldratt, *The Haystack Syndrome: Sifting Information from the Data Ocean* (North River Press, Great Barrington, MA, 1990b)

E.M. Goldratt, *The Goal: A Process of Ongoing Improvement* (North River Press, Great Barrington, MA, 1984)

E.M. Goldratt, Theo. Constraints J. **1–6** (1987)

E.M. Goldratt, *It's Not Luck* (North River Press, Great Barrington, MA, 1994)

E.M. Goldratt, *Critical Chain* (North River Press, Great Barrington, MA, 1997)

D. Lepore et al., *From Silos to Network: A New Kind of Science for Management*, SpringerBriefs in Complexity, https://doi.org/10.1007/978-3-031-40228-9

D. Kahneman, A. Tverski, *Thinking Fast and Slow* (Farrar Straus and Giroux, New York, 2013)
D. Kahneman, P. Slovic, A. Tverski (eds.), *Judgement under Uncertainty: Heuristics and Biases* (Cambridge University Press, Cambridge, 1982)
C.S. Killian, *The World of W. Edwards Deming* (SPC Press, Knoxville, Tenn, 1992)
D. Lepore, *Sechel: Logic, Language and Tools to Manage Any Organization as a Network* (Intelligent Management Inc., Toronto, 2010)
D. Lepore, *Moving the Chains: An Operational Solution for Embracing Complexity in the Digital Age* (Business Expert Press, New York, 2019)
D. Lepore, O. Cohen, *Deming and Goldratt: The Decalogue* (North River Press, Great Barrington, MA, 1999)
D. Lepore, A. Montgomery, G. Siepe, Managing complexity in organizations through a systemic network of projects, in *Applications of Systems Thinking and Soft Operations Research in Managing Complexity*. ed. by A. Masys (Springer International Publishing, Switzerland, 2016), pp. 35–69
D. Lepore, A. Montgomery, G. Siepe, *Quality, Involvement, Flow: The Systemic Organization* (CRC Press, New York, 2017)
G. Maci, D. Lepore, S. Pagano, G. Siepe, Systemic approach to management: a case study, in *Poster Presented at 5th European Conference on Complex Systems, Hebrew University, Givat Ram Campus, Jerusalem, Israel*, 14–19 September 2008
G. Maci, D. Lepore, S. Pagano, G. Siepe, Managing organizations as a system: the Novamerican case study, in *Poster Presented at International Workshop and Conference on Network Science, Norwich Research Park, UK*, 23–27 June 2008
B. Mandelbrot, *The Fractal Geometry of Nature* (W.H. Freeman, New York, 1982)
B. Mandelbrot, R.L. Hudson, *The Misbehavior of Markets: A Fractal View of Financial Turbulence* (Basic Books, New York, 2004)
H. Neave, *The Deming Dimension* (SPC Press, Knoxville, Tenn, 1990)
W.A. Shewhart, *Economic Control of Quality of Manufactured Product* (van Nostrand Company Inc., New York, 1931)
W.A Shewhart, *Statistical Method from the Viewpoint of Quality Control,* ed. by W.E. Deming (Dover, Mineola, New York, 1986)
D.J. Wheeler, *Four Possibilities* (SPC Press, Knoxville, Tenn, 1983)
D.J. Wheeler, *Understanding Statistical Process Control* (SPC Press, Knoxville, Tenn, 1992)
D.J. Wheeler, *Understanding Variation* (SPC Press, Knoxville, Tenn, 1993)
D.J. Wheeler, *Advanced Topics in Statistical Process Control* (SPC Press, Knoxville, Tenn, 1995)
D.J. Wheeler, *Building Continual Improvement* (SPC Press, Knoxville, Tenn, 1998a)
D.J. Wheeler, *Avoiding Manmade Chaos* (SPC Press, Knoxville, Tenn, 1998b)
D.J. Watt, S. Strogatz, Collective dynamics of 'small world' networks. Nature **39**, 440–442 (1998)
T. Winograd, F. Flores, *Understanding Computers and Cognition* (Addison Wesley, Boston, 1987)

Printed in the United States
by Baker & Taylor Publisher Services